Australian rainforests
Islands of green in a land of fire

Why do Australian rainforests occur as islands within the vast tracts of *Eucalyptus*? Why is fire a critical ecological factor in every Australian landscape? What were the consequences of the ice-age colonists' use of fire? In this original and challenging book, David Bowman critically examines all hypotheses that have been advanced to answer these questions. He demonstrates that fire is the most critical factor in controlling the distribution of rainforest throughout Australia. Furthermore, while Aboriginal people used fire to skilfully manage and preserve habitats he concludes that they did not significantly influence the evolution of Australia's unique flora and fauna.

This book is the first comprehensive overview of the diverse literature that attempts to solve the puzzle of the archipelago of rainforest habitats in Australia. It is essential reading for all ecologists, foresters, conservation biologists, and others interested in the biogeography and ecology of Australian rainforests.

DAVID BOWMAN is Principal Research Fellow at the Centre for Indigenous Natural and Cultural Resource Management and the School of Biological and Environmental Sciences, Northern Territory University, Darwin, Australia.

Australian rainforests

Islands of green in a land of fire

D.M.J.S. Bowman

Northern Territory University, Australia

PUBLISHED BY THE PRESS SYNDICATE OF THE UNIVERSITY OF CAMBRIDGE

The Pitt Building, Trumpington Street, Cambridge, United Kingdom

CAMBRIDGE UNIVERSITY PRESS

The Edinburgh Building, Cambridge CB2 2RU, UK www.cup.cam.ac.uk

40 West 20th Street, New York, NY 10011-4211, USA www.cup.org

10 Stamford Road, Oakleigh, Melbourne 3166, Australia

Ruiz de Alarcón 13, 28014 Madrid, Spain

First published 2000

Typeface 9.25/14pt. minion

A catalogue record for this book is available from the British Library

Library of Congress Cataloguing in Publication data

Bowman, D. M. J. S.

   Australian rainforests: islands of green in a land of fire /

D.M.J.S. Bowman.

     p. cm.

   Includes bibliographical references.

   ISBN 0 521 46568 0 (hb)

   1. Rain forest ecology – Australia.   2. Fire ecology – Australia.

   3. Fragmented landscapes – Australia.   I. Title.

QK431.B67   2000

577.34'0994–dc21   99–24978–CIP

ISBN 0 521 46568 0 hardback

Transferred to digital printing 2003

Let us regard the forests as an inheritance, given to us by nature, not to be despoiled or devastated, but to be wisely used, reverently honoured and carefully maintained.

Ferdinand von Mueller (1879)
The Chemistry of Agriculture, Melbourne.

Contents

Preface

In 1959, two papers that laid the foundations for the modern perspective on Australian rainforest ecology were published. Webb (1959) published the first systematic classification of Australian rainforest vegetation across its geographic range from Tasmania to the monsoon tropics. His project highlighted the enormous variability of Australian rainforest, notwithstanding that he excluded the transitional or ecotonal vegetation that often separates rainforests from sclerophyll vegetation from consideration. Coincidentally, Gilbert (1959) explored the dynamics of a peculiar Tasmanian forest type that was transitional between *Nothofagus cunninghamii* rainforest and tall *Eucalyptus regnans* forest. Gilbert's study was important in crystallising our perception of the importance of fire in Australian forest ecology in general and in the dynamics of *Nothofagus* rainforest boundaries in particular.

A decade later, two influential papers advanced this early work. Webb (1968) considered the environmental forces that limit the distribution of Australian rainforest. Although emphasizing a multiplicity of forces, Webb identified the fundamental importance of fire in influencing the distribution of rainforest throughout its range. In the same year, Jackson (1968) integrated a large number of ideas about the dynamic response of western Tasmanian vegetation, including *Nothofagus* rainforest, to landscape fires. Two important concepts were developed in this paper: the evolutionary relationship between fire and vegetation, and the effect on secondary successions of the variation in the interval between successive fires.

In 1969, Jones published a paper which articulated the view that Aborigines had intentionally modified their environment, including rainforest, using fire. He believed the rationale for this environmental manipulation was economic, and with a poetic flourish described it as 'fire-stick farming'. Concurrent with the emerging understanding of the dynamics of rainforest was a growing appreciation of the great antiquity of the colonisation of Australia by the Aborigines. Palynological techniques revealed that there had been massive changes in the extent of rainforest through the last glacial–interglacial cycle, raising the question of the impact of Aboriginal burning on the distribution of rainforest and the evolution of flammable non-rainforest vegetation.

Because of the enormous latitudinal and climatic range of Australian rainforest, it is inevitable that most papers have considered the ecology of rainforest at the local or regional scale. The purpose of this book is to bring together the voluminous literature on Australian rainforest in order to develop a more comprehensive explanation for the diaspora of green habitat islands in a sea of flammable forests. Such a broad geographic frame and the focus on vegetation dynamics clearly builds on the work set in train by Webb, Gilbert, Jackson, Jones and many other distinguished Australian ecologists. It is hoped that this project will complement a number of recent books. These include Adam's (1992) *Australian rainforests*, which provides a comprehensive overview of this vegetation type, Hill's (1994*a*) *History of the Australian Vegetation* which provides a comprehensive historical biogeographic context, Bond and van Wilgen's (1996) *Fire and Plants* which provides a theoretical framework for the emerging field of fire ecology, and Williams and Woinarski's (1997) *Eucalypt ecology: individuals to ecosystems*, which provides a wealth of information on the dominant vegetation on the other side of the rainforest boundary.

This task is a personal journey as much as an academic work on the environmental limits of Australian rainforest. In one way or another, I have dwelt on this problem for the last 20 years. I was first introduced to the Tasmanian form of the puzzle by Bill Jackson, who inspired me to connect my own intellectually-unstructured bushwalking experiences with his ecological theory. I was privileged to pursue some facets of the problem in my postgraduate studies, where I learnt so much from mentors including Jamie Kirkpatrick, Peter Minchin, and Mick Brown. Good fortune took me to the tropics, where Bill Freeland opened my eyes to tropical ecology and encouraged me to study fire ecology in northern Australia. Over the last decade I have been privileged to work with some exceptional biologists in the north, including John Woinarski, Peter Whitehead, Jeremy Russell-Smith, Bruce Wilson, Bill Panton, Clyde Dunlop and Rod Fensham, especially during the joint Northern Territory and Federal Government's National Rainforest Conservation Program. In the final stages of that project, Ian Gorrie organised funding for me to travel throughout Australia to visit key sites and meet key people. I owe a great debt to Len Webb and Jiro Kikkawa, who suggested that I write a book, although probably not this book, on Australian rainforest. My former employer, the Parks and Wildlife Commission of the Northern Territory, granted me leave so I could be a visiting fellow at the Australian National University and Harvard University. At the ANU I benefited from the considerable experience of Geoff Hope, Rhys Jones and John Chappell. This book could not have been written without the wonderful opportunity of a Charles Bullard Fellowship at Harvard University, where I had access to superb libraries and so many generous and helpful scholars including Peter Ashton, Fakhri Bazzaz, Barry Tomlinson, David Janos and Helene Muller-Landau. The Northern Territory University, due to Gordon Duff's adroit administration, provided financial support to allow completion of the book. I wish to thank Belinda Oliver of the Northern Territory University for tackling the herculean task of preparing the diagrams, Irene Rainey for her office assistance and Don Franklin for his patience and meticulous editorial work and proofing, and for preparing the index. Paul Adam, Dane Thomas and Rod Fensham provided valued feedback that helped improve the manuscript. My love of Australian landscapes was fostered by my parents and brother. Numerous shared adventures in the bush with my friends have been a constant inspiration for my academic research. Without Fay Johnston this project may never have been completed. Finally, I am sincerely grateful to all the other people who have generously helped me in my geographic and intellectual quest, which no doubt often appeared quixotic. Of course, the opinions, errors and omissions are my own.

1

Introduction

The existence of patches of rainforest embedded in tracts of *Eucalyptus* forests have long perplexed and sometimes astonished field biologists. The abrupt rainforest boundaries often rise up like 'a dark wall' (Figure 1.1) in the relatively open *Eucalyptus* forest (Herbert 1932) and literally confront ecologists with the question 'what determines the position of the boundaries?' The floristic differences between rainforest and *Eucalyptus* forests can be 'so great as to suggest separate geographic and historical origins in spite of their growing side by side' (Herbert 1932). The purpose of this book is to investigate the deceptively simple question of why rainforests have such limited and fragmentary coverage in Australia (Figure 1.2). This basic geographic question raises other questions such as:

(i)   Why should *Eucalyptus* and *Acacia* dominate the great bulk of Australian woody vegetation (Figure 1.3)?

(ii) How should rainforest be defined in Australia?

(iii) What environmental factors control the local extent of rainforest?

As will become apparent in this book, these questions are central issues in Australian vegetation science. The rainforest-boundary question occurs in nearly all the major arguments about Australian woody vegetation. Indeed, a number of important theories about the ecology of Australian woody vegetation have explicitly sought to resolve this problem of sharply contrasting forest types growing side by side. I recognise that the Australian rainforest boundary problem is a subset of the global biogeographic question concerning the cause of the differentiation between forest and savanna. The forest–savanna boundary is a problem of 'bewildering complexity' (Richards 1952) and has stimulated a

Figure 1.1
Abrupt subtropical rainforest boundary in Cunningham Gap, southeast Queensland. Grasstrees (Xanthorrhoea sp.) dominate the savanna in the foreground. (© Murray Fagg, Australian National Botanical Gardens.)

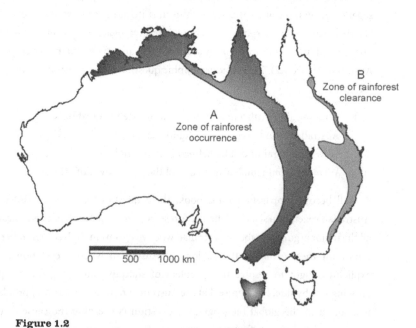

Figure 1.2
Original coverage of rainforest (closed forest and low closed forest) in Australia (A) and areas of rainforest clearance following European colonisation (B). Adapted from Anon. (1990) and Webb and Tracey (1981).

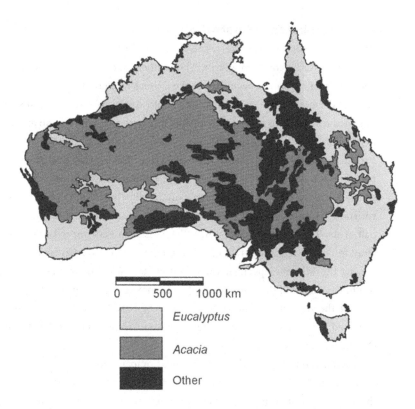

Figure 1.3

Australian vegetation dominated by Eucalyptus, Acacia or other species. Adapted from Anon. (1990). This map is Copyright © Commonwealth of Australia, AUSLIG, Australia's national mapping agency. All rights reserved. Reproduced by permission of the General Manager, Australian Surveying and Land Information Group, Department of Industry, Science and Tourism, Canberra, ACT.

sizeable literature and continuing debate as to what constitutes savanna and forest vegetation. It is outside the scope of this book to review this literature, and the reader is referred to the recent book edited by Furley *et al.* (1992). Similarly, it is not the aim of this book to provide a comprehensive overview of Australian rainforests which, in any case, is provided by the excellent book by Adam (1992). Rather, my aim is to summarise the voluminous literature about the factors controlling the distribution of rainforest in Australia and search for a general explanation of the puzzle of islands of rainforest in a sea of *Eucalyptus*-dominated vegetation.

A geographic sketch of Australia

Australia has been progressively isolated from Africa, Antarctica, India, and South America by continental drift. The southern super-continent Gondwana began to break up at the beginning of the Cretaceous Period some 150 million years (MY) before present (BP). India, South America, Africa and Australia migrated from high to low latitudes, whilst Antarctica remained in an approximately polar position (Wilford and Brown 1994). For the first half of the Tertiary Period (65 to 30 MY BP), the Australian continent lay close to Antarctica between latitudes 60 and 30 degrees south (Wilford and Brown 1994). This geographic position had a direct impact on the climate through massive annual variation in day length, with dark winters and long summer days. In the second half of the Tertiary (30 to 2 MY BP), the Australian plate moved northwards away from Antarctica at a rate of about 6 cm per year (Kershaw *et al.* 1994). During the Miocene epoch, it came into contact with the Sunda Plate (i.e. southeast Asia), and this collision resulted in mountain building on the leading edge of the Australian plate (i.e. the island of New Guinea). It also greatly increased the probability of exchanges between Australasian and Asian biotas (Raven and Axelrod 1972).

Australia is now the world's largest island. Compared to the other continents, it has the smallest land-mass (some 8×10^6 km^2), the least land above 1000 m and the most diminutive highest point (2228 m above sea level) (Jennings and Mabbutt 1986). In aggregate, Australia is the driest vegetated continent with 80% of the land area receiving less than a mean of 600 mm of rainfall per year. Nonetheless, areas with high rainfall occur in southwest Western Australia, western Tasmania and along the north and east coast of Australia (Figure 1.4). Seasonality of rainfall varies from winter rainfall in the southern part of the continent, erratic aseasonal rainfall in the centre of the continent and summer rain in the north (Figure 1.5). Extreme seasonality of rainfall occurs in the monsoon tropics, where summers are wet and winters are dry, and in southwest Western Australia where the opposite occurs. The cause of the contrasting rainfall patterns in northern and southern Australia is related to the continent's latitudinal position, lying as it does between 10° S and 44° S. In northern Australia the intertropical convergence zone is the major source of rainfall while in the south low pressure systems associated with the westerly stream are the primary source of precipitation. Mountain ranges on the east coast of Australia greatly enhance local rainfall due to orographic effects on moist air masses from the Pacific Ocean. Intense tropical low pressure systems, known in Australia as 'tropical cyclones', are an infrequent source of intense precipitation to the north

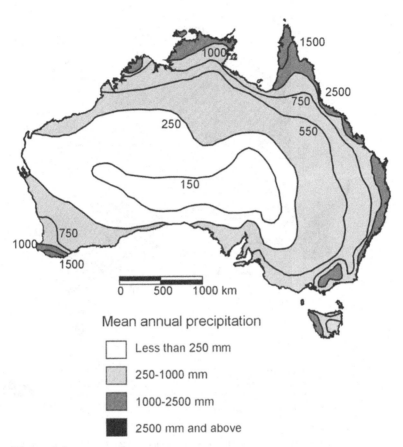

Figure 1.4
Mean annual precipitation in Australia. Adapted from Adam (1992).

of 30° S. Cyclones are an ecologically significant factor in coastal regions because of their enormously destructive winds.

The aridity of most of Australia is attributable to the belt of subtropical high-pressure systems that track across the centre of the continent during the winter and across southern Australia in the summer. These subtropical high-pressure systems produce cloudless skies and dry winds, and are responsible for the high air temperatures and great evaporative potential across much of Australia.

No month in northern Australia has an average air temperature below 18°C. Only southeastern Australia including Tasmania, a narrow coastal belt in south-western Western Australia and a montane area in central New South Wales have

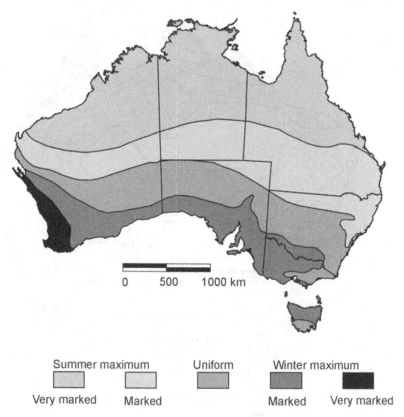

Summer maximum   Uniform   Winter maximum

Very marked   Marked   Marked   Very marked

Figure 1.5
Seasonality of rainfall in Australia. The seasonality index is calculated from the ratio of median summer (November–April) and winter (May–October) rainfall. Very marked seasonality is where maximum seasonal (summer or winter) rainfall exceeds minimum seasonal rainfall by three times and marked seasonality is where maximum seasonal rainfall exceeds minimum seasonal by 1.3 times but less than 3 times. Adapted from Anon. (1986). This map is Copyright © Commonwealth of Australia, AUSLIG, Australia's national mapping agency. All rights reserved. Reproduced by permission of the General Manager, Australian Surveying and Land Information Group, Department of Industry, Science and Tourism, Canberra, ACT.

fewer than 3 months with mean air temperature greater than 18°C (Figure 1.6). Nights with clear skies result in low nocturnal temperatures due to radiative cooling. Frosts are common in central and southern Australia, and can occur all year round in the eastern Australian mountain ranges and Tasmania (Figure 1.7).

The absence of great mountain ranges (Figure 1.8) and Australia's mid-latitudinal position permitted only localised glacial activity during the Quaternary Period (0–2 MY BP) and limits the extent of winter snow under the current

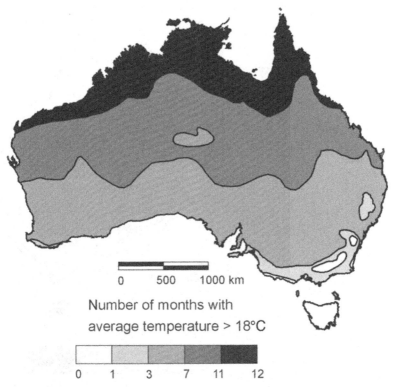

Figure 1.6
Number of months with mean air temperatures greater than 18 °C. Adapted from Anon. (1986). This map is Copyright © Commonwealth of Australia, AUSLIG, Australia's national mapping agency. All rights reserved. Reproduced by permission of the General Manager, Australian Surveying and Land Information Group, Department of Industry, Science and Tourism, Canberra, ACT.

climate. Extreme high or low temperatures are rare in Australia owing to the continent's small land mass and the moderating influence of the southern hemisphere oceans: northern hemisphere localities of similar latitude to those found in Australia experience greater temperature extremes.

Geologically, sedimentary rocks from the Precambrian Era (> 590 MY BP) and Palaeozoic Era (590–250 MY BP) dominate Australia although there are limited areas of basic igneous rocks, such as dolerite and basalt, scattered across the continent (Figure 1.9). Tertiary and Quaternary basaltic flows are restricted to the east coast where the Australian plate passed over a series of 'hotspots' (Quilty 1994). There is some evidence to suggest that the underlying geomorphology of Australian landscapes changed little during the Tertiary (Young and McDougall 1993; Nott 1995). For example, Nott (1995) suggested that many

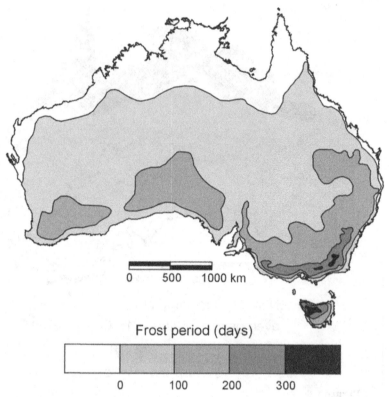

Figure 1.7
The median number of days per year when frosts are recorded. Adapted from Anon. (1986).
This map is Copyright © Commonwealth of Australia, AUSLIG, Australia's national mapping
agency. All rights reserved. Reproduced by permission of the General Manager, Australian
Surveying and Land Information Group, Department of Industry, Science and Tourism,
Canberra, ACT.

northern Australian landscapes, including 'ranges, plains and valleys', have
changed little in the last 100 MY! The stability of the land surface allowed the
production of very deeply weathered profiles (i.e. laterites to > 50 m) throughout Australia (including Tasmania, albeit where they are uncommon) (Taylor
1994). It is difficult to age lateritic surfaces, although isotopic dating of basalt
flows in north-east Queensland enabled Coventry *et al.* (1985) to demonstrate
that some lateritic profiles are 'considerably older than 6.3 MY'. Although it is
widely assumed that deep weathering reflects hot, humid conditions, this may
not be the case. Taylor (1994) noted that deep weathering might also occur
under cold, humid conditions. The dominance of tectonically stable, ancient
sedimentary rocks and very limited glaciation during the Mesozoic Era (250–65
MY BP) and Cainozoic Era (< 65 MY BP) have resulted in large areas of

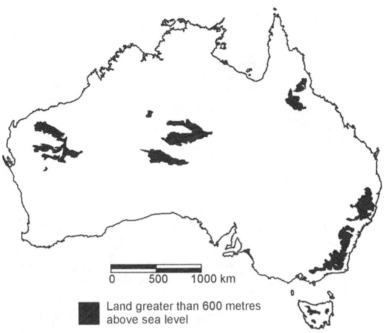

Figure 1.8
Areas greater than 600 m above sea level. Adapted from Taylor (1994) and reprinted with the permission of Cambridge University Press.

Australia having extremely old and infertile soils. Significant exceptions are fertile soils developed on basalt flows and alluvial plains.

During the Quaternary ice ages, lower sea levels exposed a vast continental shelf which increased Australia's land mass by more than 40% and united the mainland with the islands of Tasmania to the south and New Guinea to the north. These allowed the ancestors of the modern Aborigines to colonise the continent from southeast Asia about 50 000 years BP. Remarkably little is known about these ice-age Australians, indeed there is considerable debate as to the exact timing of their colonisation. Although variable across the continent, Aboriginal cultures were characteristically nomadic, tracking seasonal resource availability: no group of Aborigines practised agriculture. European colonisation occurred following the first settlements in 1788. Europeans have greatly transformed the Australian continent through intentional and unintentional introduction of numerous plants, animals and diseases, large scale clearance of native vegetation (Figures 1.2 and 1.10) and by draining wetlands, damming rivers and increasing local supplies of water by irrigation and tapping aquifers. Kirkpatrick (1994) has concisely summarised many of these profound ecological

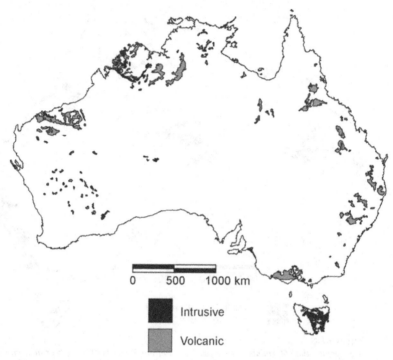

Figure 1.9
Area of basic to ultrabasic volcanic rocks (i.e. basalt, minor agglomerate, tuff) and basic to ultrabasic intrusive (dolerite, serpentine, minor norite, gabbro) in Australia, regardless of geological age. Adapted from Anon. (1988). This map is Copyright © Commonwealth of Australia, AUSLIG, Australia's national mapping agency. All rights reserved. Reproduced by permission of the General Manager, Australian Surveying and Land Information Group, Department of Industry, Science and Tourism, Canberra, ACT.

transformations. These changes are ongoing, but have had little direct effect on large areas of northern Australia. There European population densities remain very low and Aboriginal populations are relatively high. Nonetheless, the breakdown of Aboriginal fire-management practices, weed invasion, decline of some mammal and bird species and damage by feral and domesticated stock suggest that the apparently intact north Australian ecosystems have been degraded since European colonisation.

Geographic pattern of Australian rainforest

Putting aside the vexatious issue of rainforest definition for the moment, I shall briefly describe the distribution patterns of Australian rainforest at the continental, regional and local scales. Australian rainforests have a fragmentary distribu-

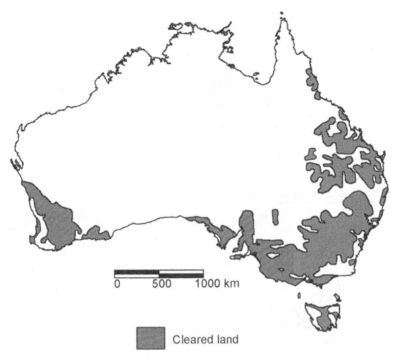

Cleared land

Figure 1.10
Areas of native vegetation cleared between 1788 and 1980. Adapted from Lowe (1996), Commonwealth of Australia copyright reproduced by permission.

tion across a coastal arc of some 6000 km from the Kimberley in Western Australia to southwest Tasmania (Figure 1.2). The inland limit of rainforest occurs at approximately the 600 mm isohyet. Curiously, the high rainfall region of southwestern Western Australia does not support any rainforest vegetation. Continental and regional scale maps of rainforest mask the high degree of rainforest fragmentation, which is more accurately illustrated by detailed vegetation maps of four regions across the geographic range of Australian rainforest (Figure 1.11).

On the east coast, some large tracts of rainforest are associated with fertile soils derived from basalt flows, although there is by no means a perfect correlation between soil fertility and rainforest distribution (Figures 1.2 and 1.9). For example, large tracts of rainforest occur in southwestern Tasmania and Cape York Peninsula on infertile soils derived from quartzite rocks and granites respectively.

Not only does Australian rainforest have a wide latitudinal range from southern Tasmania to northern Australia; it also has a very wide altitudinal range.

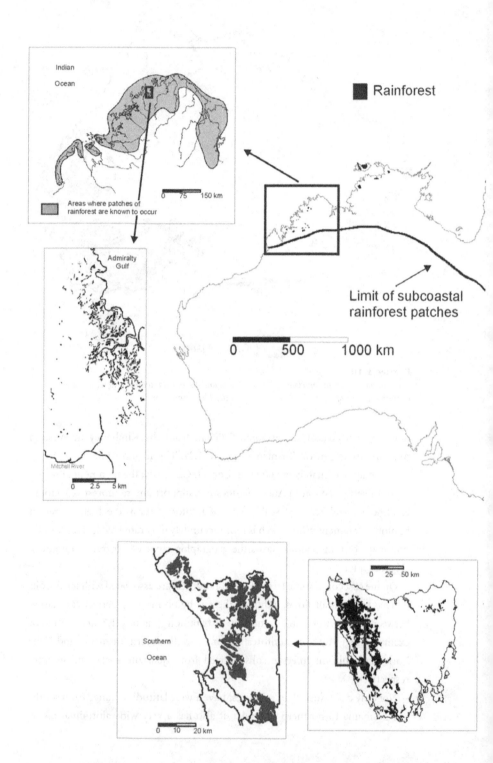

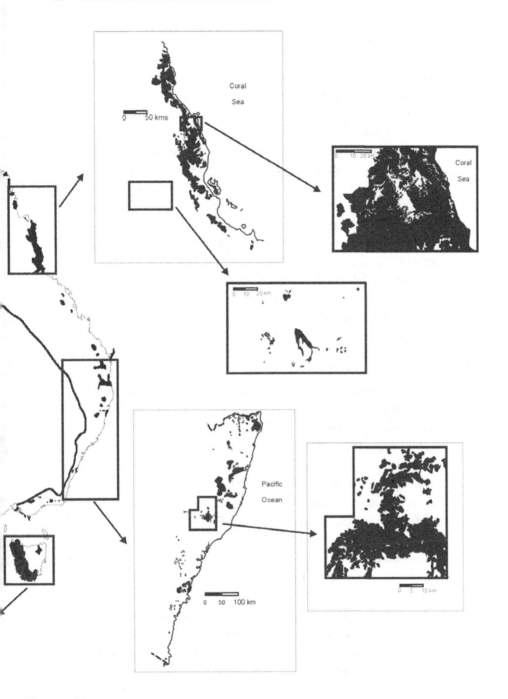

Figure 1.11
Maps showing the effect of spatial scale on the distribution of rainforest in four regions across the arc of rainforest in Australia. Adapted from Adam (1992), Kirkpatrick (1977), Dodson and Myers (1986), Ash (1988), and Clayton-Greene and Beard (1985).

Throughout its range, rainforest occurs from sea level to the altitudinal limit of woody vegetation; indeed some types of rainforest are restricted to high elevation sites (e.g. montane tropical rainforests, and alpine rainforests in Tasmania).

Webb *et al.* (1984) undertook a numeric biogeographical analysis of Australian rainforests based on floristic lists collected across the geographic range of the vegetation. They identified three broad classificatory groups of rainforest that correspond to climatic zones which they described as 'ecofloristic regions': their 'Group A' was mainly temperate but includes some areas in the humid subtropics; 'Group B' was characterised by tropical forests; and 'Group C' was characterised by rainforests in the subtropics. However, within these three broad groups Webb *et al.* (1984) identified a total of eight further subdivisions which they described as 'ecofloristic provinces' (see Table 1.1 and Figure 1.12). Only three (A_3; B_3; C_2) of these eight groups were geographically distinct (Figures 1.13 and 1.14). The other groups, which all occur on the east coast, overlap to greater or less amounts (Figure 1.15). They explained the overlap between the various 'ecofloristic provinces' as being a consequence of habitat heterogeneity in eastern Australia associated with steep altitudinal and rainfall gradients and the juxtaposition of sites with contrasting geology and hence soil fertility. They also suggested that some of the overlaps might be explained as the result of long-term climatic changes that have resulted in 'relictual' or environmentally-discordant distributions of some rainforest types. Clearly, the complex distribution patterns of rainforest types prohibit the rigid application of simple climatic classifications such as tropical and subtropical rainforest. In any case, Webb *et al.* (1984) demonstrated that Australian rainforests can legitimately be seen as a floristic continuum with a considerable number of species and genera co-occurring between 'ecofloristic regions' (Figure 1.16). For example, about 13% of the rainforest tree species occur in all three 'ecofloristic regions'. Thus any biogeographic classification of Australian rainforests is necessarily imperfect, and definitions of regions ultimately include an arbitrary component. In this book, I have occasionally broken up the huge continuum of Australian rainforests into four geographic zones: monsoon tropics, humid tropics, subtropics and temperate. I acknowledge that this division ignores the complexity of altitudinal effects on climate and edaphic influences on rainforest vegetation. I have pursued this classification because it is a convenient method of relating studies that have been conducted in the same geographic zone, and in some cases, the same landscape.

The transition between rainforest and the surrounding non-rainforest vegetation is typically abrupt for monsoon and tropical rainforests (Figure 1.17). However, in some situations rainforest invades the surrounding *Eucalyptus*

Figure 1.12
Boundaries of 'ecofloristic provinces' defined by Webb et al. (1984).

TABLE 1.1. Climatic variables for characteristic meteorological stations for eight major rainforest 'ecofloristic provinces' in Australia

Ecofloristic province	Climate type	Latitude and longitude	Altitude	Mean annual rainfall (mm)	Mean annual rain days	Mean rainfall in driest 6 consecutive months (mm)	Mean annual air temperature (°C)	Mean minimum temperature of coldest month (°C)
A₁	aseasonal humid warm subtropical	28° 19' S 153° 26' E	5	1722	142	550	19.3	5.8
	subtropical	30° 19' S 153° 07' E	3	1759	147	569	18.3	6.6
	aseasonal humid cool subtropical	28° 36' S 153° 23' E	381	2388	147	762	16.6	5.8
	subtropical	31° 23' S 152° 15' E	146	1603	159	499	17.3	3.8
A₂	aseasonal humid	33° 42' S 150° 22' E	883	1374	149	406	12.3	1.5
	montane subtropical	30° 37' S 152° 11' E	1036	1516	130	444	13.1	2.8
	aseasonal humid warm temperate	37° 32' S 149° 09' E	88	1004	146	448	14.1	2.0

		Coordinates							
A₃	aseasonal humid cool temperate	41° 26' S	145° 31' E	624	2201	252	820	8.3	1.6
		41° 38' S	145° 57' E	915	2774	237	1066	6.7	0.0
B₁	seasonal humid tropical	12° 26' S	130° 52' E	29	1594	109	110	27.5	18.9
		12° 31' S	138° 03' E	7	1360	92	51	27.8	17.9
		12° 47' S	143° 18' E	19	2049	202	215	25.4	18.4
		12° 27' S	142° 38' E	39	1362	102	66	26.2	17.0
B₂	aseasonal humid tropical	17° 32' S	145° 58' E	40	3644	155	760	23.5	15.1
		18° 16' S	146° 02' E	5	2127	122	289	24.1	13.3
		17° 12' S	145° 34' E	715	1260	113	189	20.2	10.8
B₃	seasonal subhumid tropical	13° 57' S	143° 12' E	193	1146	86	44	25.4	16.7
		15° 28' S	141° 25' E	13	1222	71	36	26.9	14.8
		14° 28' S	132° 16' E	107	952	62	46	27.2	12.9

TABLE 1.1. (*continued*)

Ecofloristic province	Climate type	Latitude and longitude	Altitude	Mean annual rainfall (mm)	Mean annual rain days	Mean rainfall in driest 6 consecutive months (mm)	Mean annual air temperature (°C)	Mean minimum temperature of coldest month (°C)
C_1	seasonal humid warm subtropical	27° 28' S 153° 02' E	38	1146	123	366	20.7	9.8
		26° 11' S 152° 40' E	94	1161	117	354	20.4	6.0
		24° 42' S 151° 18' E	339	905	89	245	18.7	3.7
C_2	seasonal subhumid warm subtropical	24° 24' S 150° 30' E	173	699	75	187	20.7	5.1
		18° 09' S 144° 19' E	453	799	57	67	23.8	9.5
		26° 30' S 147° 59' E	335	575	61	188	19.8	3.3

Adapted from Webb *et al.* (1984).

Figure 1.13
Interior of a Nothofagus cunninghamii *rainforest on fertile soil in northwestern Tasmania. This forest type is typical of the rainforests in the A₃ ecofloristic province (Webb et al. 1984). (Photograph: David Bowman.)*

Figure 1.14
Boundary of a monsoon vine-thicket rainforest on basalt soils in northeastern Queensland. This forest type is typical of rainforest in the B₃ ecofloristic province (Webb et al. 1984). (Photograph: David Bowman.)

Figure 1.15
Profile of humid tropical rainforest near Cape Tribulation in northeastern Queensland. This forest type is typical of rainforest in the B_1 ecofloristic province (Webb et al. 1984). (© Murray Fagg, Australian National Botanical Gardens.)

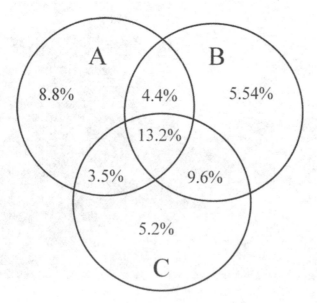

Figure 1.16
Percentage of rainforest species from a sample of 1316 tree species that occur in the three 'ecofloristic regions' that divide up the geographic range of Australian rainforest. Adapted from Webb et al. (1984).

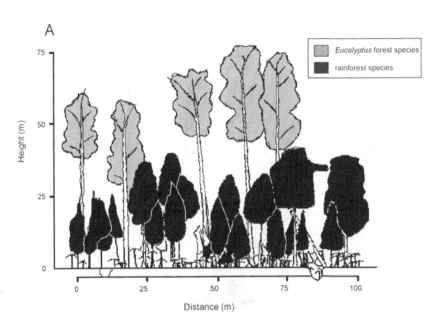

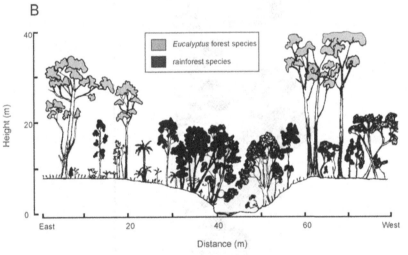

Figure 1.17

Profile diagrams of rainforest boundaries from four regions in Australia.
A. Mixed Eucalyptus–Nothofagus cunninghamii *rainforest in Tasmania (adapted from*
Gilbert 1959). B. Riverine rainforest dominated by Acmena smithii *and* Tristaniopsis laurina
within a Eucalyptus *forest in Victoria (adapted from Melick and Ashton 1991). C. Humid*
tropical rainforest boundary on the Atherton Tablelands in northeastern Queensland (adapted
*from Unwin 1989). D. Monsoon rainforest–*Eucalyptus *savanna boundary on Groote Eylandt,*
Northern Territory (adapted from Langkamp et al. 1981, with kind permission from Kluwer
Academic Publishers).

C and D overleaf

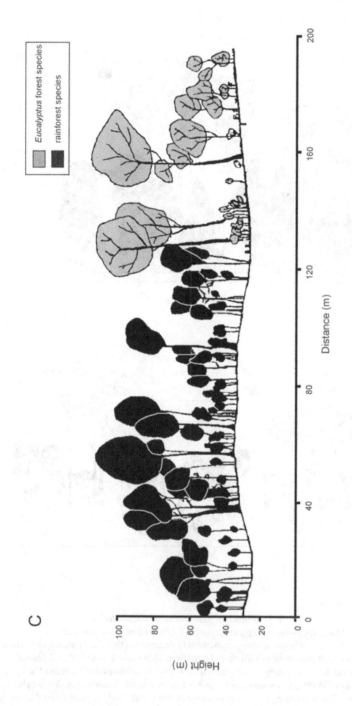

C

Height (m)

Distance (m)

Eucalyptus forest species

rainforest species

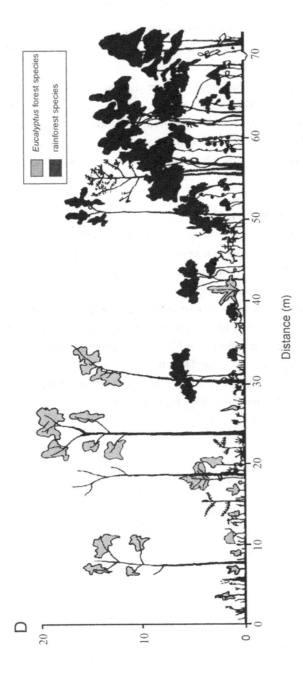

D

20

10

0

Distance (m)

0   10   20   30   40   50   60   70

Eucalyptus forest species

rainforest species

forests (Figure 1.17). The width of these rainforest ecotones is highly variable but typically less than one km. The boundary of temperate rainforests can be either abrupt or extremely gradual. Wide ecotones are often recognised as a distinct community known as 'mixed forests' (Figure 1.17), and in Tasmania there are large tracts of these mixed forests. The mixture of rainforest and *Eucalyptus* forest has bedevilled neat definitions of 'rainforest' in Australia. The definition of rainforest in Australia is the subject of the following two chapters. The remainder of the book seeks to explain why Australian rainforests have a fragmentary distribution.

Fire, air, earth and water

A number of theories have been advanced to explain the control of rainforest in Australia. Some of these theories have been concealed, Trojan-horse-like, within definitions of 'rainforest'. Other theories are explicit and typically have championed the primacy of single environmental contingencies such as fire history, environmental changes, soil fertility, or water stress. Some theories are so broad in scope that they cover all bases by acknowledging the importance of many factors and their complex interactions. Clearly, the various theories are competing for the same intellectual space. The aim of this book is to subject them to critical analysis to determine if there is a general explanation for the continent-wide fragmentary distribution of Australian rainforest.

2

What is Australian rainforest?

The Europeans who colonised Australia had no conceptual frame with which to deal with the vast tracts of evergreen native vegetation that they came to, at first disparagingly, then affectionately, call 'bush'. The place name 'Botany Bay' attests to the impact that the 'strange' Australian flora had on the earlier explorers. In the nineteenth century, the dominant Australian vegetation aroused emotions of alienation, fear, despair, desolation, and melancholia in many observers. For example, the author Marcus Clarke wrote in 1876 that 'The Australian forests are funereal, secret, stern. Their solitude is desolation... No tender sentiment is nourished in their shade...' (Ritchie 1989). One exception was the luxuriant, cool and shaded forests that formed sanctuaries in the vast tracts of dry 'monotonous' bush (Ritchie 1989). In the nineteenth century these moist forests were known, not as 'rainforests' but by a range of terms including fern-forest, brush, scrub, and jungle (Adam 1992). In the twentieth century, the term 'rainforest' has been almost universally used and understood to denote an atypical Australian forest type. Sadly, we have limited appreciation of Aboriginal knowledge or terminology of Australian vegetation: only a few Aboriginal terms for Australian plant communities, such as 'mulga' and 'mallee', have been incorporated into the English language.

Although a number of complex taxonomic systems have been developed to classify variation amongst rainforest types (e.g. Webb 1959, 1968, 1978; Jarman *et al.* 1987; Floyd 1990; Russell-Smith 1991; Fensham 1995), there has been little agreement as to the bounds of the concept of rainforest in Australia. Indeed Webb and Tracey (1981) note that scientific definition of Australian rainforest has become 'increasingly elusive' despite ongoing research. The purpose of this chapter is to explore the concept of Australian rainforest.

Classificatory systems of Australian rainforest
Schimper and the first definition of rainforest in Australia

The term 'rainforest' is a translation of 'Regenwald', which was coined by the German botanist Schimper. It first appears in his monumental book *Plant-Geography upon a Physiological Basis*, which was published in German in 1898 and in English in 1903. In this book, Schimper developed a general theory of the climatic control of vegetation formations on Earth. In brief, he argued that increasing length of seasonal drought in the tropics resulted in the following sequence of tropical vegetation formations: rainforest; monsoon forest; tropical grassland; savanna–forest; thorn forest and desert scrub. Schimper had limited climatological data with which to define the climates of these six formations, although he suggested that tropical rainforest would occur in environments with at least 180 cm of annual rainfall and no seasonal moisture deficit (Figure 2.1).

Unlike the tropics, Schimper (1903) believed that the duration and temperature of seasonal drought (i.e. winter versus summer) was of critical importance in controlling vegetation formations in the temperate zone. He argued that, in temperate climates with mild winters and summer rainfall, impoverished rainforest formations would develop, whereas in temperate climates with mild winters and summer drought the development of a temperate savanna forest would be favoured. He believed there was no tropical analogue for temperate savanna forest, which he described as 'sclerophyllous forests' because of the dominance of plants with thick, leathery or sclerophyllous leaves. Schimper classified most temperate *Eucalyptus* forests as belonging to his sclerophyll savanna–forest formation. He wrote that *Eucalyptus* 'forest, though it differs to some extent from ordinary savannah–forest, for instance, frequently though not always, in the greater height of its trees and in its evergreen foliage, does not differ in essentials, for instance in the rich growth of grass between the trunks that rise out of the very open wood and in the absence or very feeble development of the underwood'. However, Schimper did not classify all temperate *Eucalyptus* communities as sclerophyll forest. Tall *Eucalyptus* forests of Tasmania and Victoria conformed to his definition of temperate rainforest because they occurred in environments with high summer rainfall (Figures 2.2 and 2.3).

Rainforest defined by climatological parameters

Since Schimper there have been a number of other attempts to classify global vegetation formations on the basis of climatological parameters, for example Holdridge (1947). Although Richards (1952) pointed out in his classic text on

Figure 2.1
The interior of an archetypal humid tropical rainforest at Mission Beach near Tully in north Queensland. (© Murray Fagg, Australian National Botanical Gardens.)

tropical rainforests that simple climatic definitions are unrealistic given the local importance of edaphic factors and fire in influencing vegetation, he and many other tropical ecologists have been unable to offer a substantially improved definition of rainforest (e.g. Whitmore 1990). For instance, the recent atlas of the tropical forests in Asia and the Pacific (including Australia) used a modified version of Schimper's scheme (Collins *et al.* 1991). These authors defined tropical rainforest as occurring in climates where there are no more than two months that receive less than 60 mm of rainfall, while monsoon forest occurs where there are at least three months with less than 60 mm of rainfall.

Rainforest defined by a priori description

An alternative strategy to define rainforest has been to produce an a priori description of its physiognomy. To some extent Schimper (1903) also did this when he gave his subsequently much-quoted description of tropical rainforest as 'evergreen, hygrophilous in character, at least thirty metres high, but usually taller, rich in thick-stemmed lianes, and in woody as well as herbaceous epiphytes' (Figure 2.4). This definition was not rigid, however, as Schimper realised that the rainforest is less luxuriant in subtropical and temperate zones. His definition of temperate rainforest still focused on the occurrence of 'evergreen

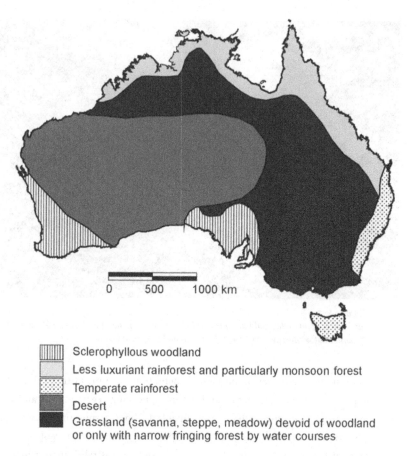

Sclerophyllous woodland
Less luxuriant rainforest and particularly monsoon forest
Temperate rainforest
Desert
Grassland (savanna, steppe, meadow) devoid of woodland
or only with narrow fringing forest by water courses

Figure 2.2
Distribution of major Australian vegetation formations including rainforest according to Schimper (1903).

hygrophilous trees' but allowed the occurrence of summer-green sub-dominants and emphasised the simplification of physiognomic features relative to tropical rainforest. Examples of the latter were smaller and firmer leaves, smaller and less diverse lianes, and the rarity of trees with plank buttresses.

   Beadle and Costin (1952) developed the first comprehensive a priori classification of Australian vegetation. At the formation level, Beadle and Costin recognised six arborescent vegetation formations: savanna, shrub, mallee, woodland, sclerophyll forest and rainforest. Their definition of Australian rainforest is as follows: 'a closed community dominated by usually mesomorphic meso- or megaphanerophytes forming a deep densely interlacing canopy in which lianes and epiphytes are invariably present, with mesomorphic subordinate strata of

Figure 2.3
Tall Eucalyptus regnans
forest in a high rainfall
area of Victoria. Schimper
classified these forests as a
form of temperate
'rainforest'. (Photograph:
David Bowman.)

smaller trees, shrubs, and ferns and herbs'. Their classification was a purely
descriptive device that explicitly sought to avoid the 'erroneous impression that
climate is the controlling factor in determining vegetation' although they used
climatic terms (temperate, subtropical, tropical and monsoonal) to define four
rainforest sub-formations. They also used the terms 'wet' and 'dry' to denote the
degree of mesomorphism of some non-rainforest communities and this aspect of
their system has been widely adopted, as evidenced by the frequent Australian
usage of the terms 'wet sclerophyll forest' and 'dry sclerophyll forest'.

Baur (1957, 1965) believed that the Beadle and Costin (1952) definition of
rainforest 'excluded communities that are characteristically regarded as being
rain-forest'. He proposed the following a priori definition for rainforests in the
State of New South Wales: 'a closed, moisture-loving community of trees,
usually containing one or more subordinate storeys of trees and shrubs; frequently mixed in composition; the species typically but not invariably broad-

Figure 2.4
A subtropical rainforest,
Lamington National Park,
southeast Queensland, with
conspicuous woody vines.
(Photograph: Don
Franklin.)

leaved and evergreen; heavy vines (lianes), vascular and nonvascular epiphytes, stranglers and buttressing often present and sometimes abundant; floristic affinities mainly with the relic Gondwanan flora; eucalypts typically absent except as relics of an earlier community'. He recognised all of Beadle and Costin's (1952) four rainforest sub-formations (temperate, subtropical, tropical and monsoonal) but suggested that rather than using the term 'monsoon rainforest' it was more appropriate to use the oxymoronic term 'dry rainforest'. He justified this by noting that his conception of dry rainforests bore little resemblance to the monsoon rainforest originally described by Schimper (1903). The term 'dry rainforest' was not without precedent, having been coined by Beard (1955) to describe a rainforest formation in the central American tropics that occurred on very well-drained sandy soils. Floyd (1990) has adopted a slight modification of Baur's scheme in his recent treatment of the rainforests of New South Wales.

The UNESCO (1973) a priori description of global vegetation types is based

Figure 2.5
The buttressed trunk of Ficus racemosa *gives a compelling sense that monsoon rainforests are an attenuated form of 'true' tropical rainforest. (Photograph: David Bowman.)*

on the physiognomy and structure of vegetation and is philosophically similar to the descriptive classification proposed by Beadle and Costin (1952) and Baur (1965). The UNESCO system is not popular with Australian vegetation scientists, possibly because its titles for vegetation formations are opaque and unfamiliar to Australian ecologists. For instance a tropical rainforest is described as a 'closed, mainly evergreen, tropical ombrophilous forest' and tall *Eucalyptus* forest is described as 'winter-rain evergreen sclerophyllous lowland and submontane forest'. Carnahan (1981) considered that the UNESCO system did not deal adequately with Australia's unique flora, although he did not expand upon this point.

Rainforest defined by diagnostic life forms

Following Schimper, a number of Australian authors have defined rainforest according to the presence or absence of various plant life forms such as epiphytes, vines, and trees with buttresses (Figure 2.5). For example, in the context of subtropical and tropical rainforests, Herbert (1960) suggested that forests that lacked epiphytes could not be considered 'true rainforest'. Francis (1951) considered that an important diagnostic feature of Australian rainforest was the 'very noticeable suppression of the special adaptations to dry conditions such as are evident in the open *Eucalyptus* and *Acacia* forests'. These adaptations include tolerance of bushfires, and hard-textured narrow leaves that expose a minimum area to direct sunlight. The absence of grass is often seen as a basic feature of rainforest, although moist *Eucalyptus*, or wet sclerophyll, forests also lack grass cover (e.g. Ashton 1981*b*). Webb (1959) provided the following general description of rainforest as being a floristically complex closed forest with closely spaced trees arranged in several strata and containing a prominence of specialised life forms such as epiphytes and lianes, and absences of annual herbs. However, he recognised that there was enormous variation in the life form spectra and physiognomy of rainforest (Webb 1959, 1968, 1978). He developed a key (Table 2.1), widely employed by Australian ecologists, that is based on the following four categories of physiognomic and structural features:

(i)   canopy characteristics including height, canopy evenness, density of emergents, crown shapes and number of strata;

(ii)   leaf size, shape and texture and the degree and timing of leaf-fall;

(iii) bark texture and colour, epiphyte load, occurrence of cauliflory, buttressed trunks, surface root mats and stilt roots;

(iv)   occurrence of specialised growth forms such as palms, tree ferns, woody and wiry vines and gigantic herbs such as gingers, bananas and bamboos.

Webb freely acknowledged his classification was imperfect, and subsequently provided a detailed list of physiognomic and structural characters that could be used in numerical analysis in order to objectively classify rainforest vegetation (Webb *et al.* 1976).

Rainforest defined by canopy measurement

A quantitative method of classifying Australian vegetation, including rainforest, was developed by Wood and Williams (1960) in a two-way table comprising four classes of canopy cover (very dense, dense, mid-dense, open) and the dominance of four life forms (trees, tall shrubs, low shrubs and herbs). In this system, rainforests were classified as communities of trees, tall and low shrubs that form dense canopies, where the distance between crowns is less than the diameter of a tree crown. Specht (1981*a*) refined this system by developing a two-way table with five categories of foliage projective cover of the tallest stratum and a variety of life forms (Table 2.2). Specht chose to avoid the term 'rainforest', proposing the synonym 'closed forest' instead. According to Specht's scheme, 'closed forest' is vegetation where the tallest stratum is composed of trees and the foliage of the tallest stratum covers more than 70% of the view of the sky when viewed

TABLE 2.1. Field key to structural types of Australian rainforest vegetation according to Webb (1978)

1.	Mesophylls and notophylls most common	2
	Notophylls and microphylls most common	7
	Microphylls most common	10
	Nanophylls most common	13
	(Leaf lengths: mesophyll: 12.5–25 cm; notophyll: 7.5–12.5 cm; microphyll: 2.5–7.5 cm; nanophyll < 2.5 cm)	
2.	Robust lianes, vascular epiphytes, plank buttresses, macrophylls (> 25 cm leaf length) and compound mesophylls prominent; trunk spaces generally obscured by aroids and palms; stem diameters irregular, many av. 60–120 cm; canopy level av. 21–42 m	3
	Robust lianes and vascular epiphytes not conspicuous in upper tree layers which are simplified; spur rather than plank buttresses prominent; trunk spaces open, stem diameters (except for evergreen emergents) generally regular, av. 60 cm; canopy level av. 24–36 m. Simplification of structural features does not, however, approach that of simple notophyll evergreen types. Sclerophylls (e.g. *Acacia*) may be scattered in canopy	6
3.	Deciduous emergent and top canopy trees rare	4
	Deciduous and semi-deciduous emergent and top canopy	5
4.	Palm trees not prominent in canopy	**Complex mesophyll vine forest (CMVF)**
		Tropical rainforest
	Feather palm trees prominent in canopy	**Mesophyll feather-palm vine forest (MFPVF)**
		Tropical rainforest

TABLE 2.1. (*continued*)

5.	Mostly mesophylls	**Semi-deciduous mesophyll vine forest (SDMVF)**
		Tall monsoon forest, wet monsoon forest
	Mostly notophylls	**Semi-deciduous notophyll vine forest (SDNVF)**
		Monsoon forest, wet monsoon forest
6.	Palm trees not prominent in canopy	**Mesophyll vine forest (MVF)**
		Tropical rainforest, wet monsoon forest
	Fan palm trees prominent in canopy	**Mesophyll fan-palm vine forest (MFAPVF)**
		Tropical rainforest, wet monsoon forest
7.	Robust and slender woody lianes, vascular epiphytes, plank buttresses, and compound entire leaves prominent; trunk spaces generally obscured by the aroid *Pothos*; stem diameters irregular, many av. 60–120 cm	8
	Robust lianes and vascular epiphytes inconspicuous in tree tops; slender woody and wiry lianes prominent in understorey; plank buttresses inconspicuous; simple toothed leaves prominent; trunk spaces open; stem diameters (except for emergents) generally regular av. 60 cm; tree crowns evergreen and generally sparse and narrow; strong tendency to single species dominance (e.g. *Ceratopetalum*) in upper tree layers; canopy level even, av. 21–33 m often with sclerophyllous emergents and co-dominants	**Simple notophyll evergreen vine forest (SNEVF)**
		Subtropical rainforest, warm temperate rainforest, coachwood rainforest
	Robust lianes, vascular epiphytes and plank buttresses present, but not so prominent as in complex types; tree crowns mostly evergreen, but with a few semi-evergreen or deciduous species, i.e. structural features are intermediate between simple and complex types	**Notophyll vine forest (NVF)**
		Subtropical rainforest

Robust and slender lianes generally present, wiry lianes (climbing ferns) generally conspicuous; feather palms generally conspicuous; tree crown evergreen; canopy level av. 20–25 m

Evergreen notophyll vine forest (ENVF) ± feather palms

Monsoon rainforest, wet monsoon forest

Robust, slender and wiry lianes generally inconspicuous; fleshy vascular epiphytes may be prominent on trunks; plank buttresses inconspicuous; simple entire leaves prominent; deciduous species generally absent but many tree crowns become sparse during the dry season, i.e. semi-evergreen; typically mixed with sclerophyllous emergents and co-dominants 9

8. Canopy level uneven, av. 21–45 m, emergents mostly evergreen and umbrageous

Complex notophyll vine forest (CNVF)

Subtropical rainforest

Canopy level uneven, av. 15–36 m, occasional deciduous species with common emergent *Araucaria* or *Agathis*, reaching av. 36–51 m

Araucarian notophyll vine forest (ANVF)

Hoop pine rainforest

9. Canopy level av. 10–20 m

Simple semi-evergreen notophyll vine forest (SSENVF)

Monsoon forest, dry rainforest, dry monsoon forest

Canopy level av. 3–9 m, generally even, and canopy trees often branched low down (shrub-like)

Simple semi-evergreen notophyll vine thicket (SSENVT)

Monsoon forest, dry rainforest, dry monsoon rainforest

TABLE 2.1 (*continued*)

10. Mossy and vascular epiphytes inconspicuous in top tree layers; robust lianes generally prominent; plank buttresses absent; prickly and thorny species frequent in usually dense shrub understorey; ground layer sparse; compound leaves and entire leaf margins common 11

    Mossy and vascular epiphytes usually present in top tree layers; robust lianes inconspicuous; slender and wiry lianes generally prominent; plank buttresses absent; prickly and thorny species absent; simple leaves with toothed margins common; strong tendency to single species dominance (*Nothofagus, Eucryphia*) in tree layer; tree ferns and ground ferns prominent; sclerophyll emergents generally present in marginal situations 12

11. Canopy level uneven, av. 9–15 m with mixed evergreen and semi-evergreen emergent and upper tree layer species; araucarian and deciduous emergents rare or absent

    Low microphyll vine forest (LMVF)

    Dry rainforest, dry monsoon forest

    Canopy level uneven, av. 9–15 m with some deciduous and semi-evergreen species; frequent araucarian (*Araucaria cunninghamii*) emergents to av. 21–36 m

    Araucarian microphyll vine forest (AMVF)

    Hoop pine rainforest, dry rainforest

    Canopy level uneven, av. 4–9 m with mixed evergreen, semi-evergreen and deciduous emergents to av. 9–18 m, swollen stems ('Bottle Trees') common

    Semi-evergreen vine thicket (SEVT)

    Bottle-tree scrub, dry monsoon rainforest, dry rainforest

Canopy level uneven and discontinuous, av. 4–9 m; practically all emergents are deciduous, and many understorey species are deciduous or semi-evergreen; swollen stems ('Bottle Trees' and other species) may be common

Deciduous vine thicket (DVT)

Monsoon forest, dry monsoon forest, dry rainforest

12. Canopy level tall, even except for sclerophylls, av. 20–45 m

Microphyll fern forest (MFF)

Warm temperate rainforest, Beech forest

Canopy level stunted, generally even and mixed with sclerophylls, av. 6–9 m

Microphyll fern thicket (MFT)

Warm temperate rainforest, Beech forest

13. Canopy level tall, except for sclerophylls, av. 18–40 m

Nanophyll fern forest (NFF) and mossy forest (NMF)

Cool temperate rainforest, Beech forest, Myrtle forest

Canopy level stunted, uneven, often with sclerophylls, av. 6–9 m

Nanophyll fern thicket (NFT) and mossy thicket (NMT)

Cool temperate rainforest, Beech forest, Myrtle forest

Synonyms of rainforest types are in italics, and are adapted from Webb (1978), Winter *et al.* (1987), Bowman *et al.* (1991*a*), Russell-Smith (1991) and Fensham (1995).

TABLE 2.2. Structural formations of Australian vegetation

Life form of tallest stratum		Foliage projective cover of tallest stratum				
		100–70% (4)[b]	70–50% (3+)	50–30% (3–)	30–10% (2)	> 10% (1)
Trees[a] > 30 m	(T)[b]	Tall closed forest	Tall forest	(Tall open forest)[c]	(Tall woodland)[c]	—
Trees 10–30 m	(M)	Closed forest	Forest	Open forest	Woodland	Open woodland
Trees < 10 m	(L)	Low closed forest	Low forest	Low open forest	Low woodland	Low open woodland
Shrubs[a] > 2 m	(S)	Closed scrub	Scrub	Open scrub	Tall shrubland	Tall open shrubland
Shrubs 0.25–2 m Sclerophyllous	(Z)	Closed heathland	Heathland	Open heathland	Shrubland	Open shrubland
Non-sclerophyllous	(C)	—	—	Low shrubland	Low shrubland	Low open shrubland
Shrubs < 0.25 m Sclerophyllous	(D)	—	—	—	Dwarf open heathland (fell-field)	Dwarf open heathland (fell-field)

					Dwarf shrubland	Dwarf open shrubland
Non-sclerophyllous	(W)	—	—	—	—	—
Hummock grasses	(H)	—	—	—	Hummock grassland	Open hummock grassland
Herbaceous layer						
Graminoids	(G)	Closed (tussock) grassland	(Tussock) grassland	(Tussock) grassland	Open (tussock) grassland	Very open (tussock) grassland
Sedges	(Y)	Closed sedgeland	Sedgeland	Sedgeland	Open sedgeland	Very open sedgeland
Herbs	(X)	Closed herbland	Herbland	Herbland	Open herbland	Very open herbland
Ferns	(F)	Closed fernland	Fernland	Fernland	—	—

[a] A tree is defined as a woody plant usually with a single stem; a shrub is a woody plant usually with many stems arising at or near the base.

[b] Symbols and numbers given in parentheses may be used to describe the formation, e.g. tall closed-forest: T4, hummock grassland: H2.

[c] Senescent phases of Tall forest.

Modified from Specht (1981a).

vertically from the forest floor. Trees are defined as a 'woody plant usually with a single stem'. The closed forest category is broken up into three subdivisions depending on canopy height: > 30 m form 'tall closed forests'; 10–30 m form 'closed forests'; 2 m and < 10 m form 'low closed-forests'. A defect in Specht's scheme is that there is no regard to the floristic composition of the vegetation: a closed forest may be a mangrove forest, a rainforest or mixtures of rainforest species and species such as *Eucalyptus* not thought to be typical of rainforest (Jarman and Brown 1983). In an attempt to resolve the problem, Walker and Hopkins (1984) suggested that rainforest as defined by Webb (1978) should be treated as a special category of Specht's (1981*a*) concept of 'closed forests'. Specht's (1981*a*) scheme is not neutral regarding environmental parameters, however. Specht (1981*b*) argued that his system is based upon eco-physiological principles which demonstrate that closed forests only occur were there is near optimal soil moisture. In a sense Specht's argument is a recasting of Schimper's (1903) original theory that climate controls vegetation. Specht's theory will be considered in detail in Chapter 6.

Rainforest defined by light environment

An obvious feature of boundaries between rainforests and adjacent non-rainforest formations is a dramatic change in microclimate, and particularly the light environment. Seddon (1984) noted that light regime is often used to dichotomise these formations. He wrote: 'The tendency in Australia for botanists in their weaker moments to use 'closed-forest' as if it were synonymous with 'rainforest' really indicates rather clearly that this is the only forest form that is at all common in Australia that *does* cast a dense shade. It is so different from the familiar, light-drenched open forest dominated by eucalypts'. It has been widely assumed that the differences in the light environment between rainforest and non-rainforest formations are of significance in controlling the establishment of tree species within the rainforest (e.g. Herbert 1932; Francis 1951; Jackson 1968). The light environment in Australian rainforests and the photosynthetic response of rainforest and non-rainforest to high and low light conditions is considered in Chapter 7.

Rainforest defined by biogeographically distinctive taxa

In addition to the structural and physiognomic features of Australian rainforests, many early workers were struck by the floristic differences between typical Australian vegetation and Australian rainforests. A widely held explanation for the distinct floristic composition of *Eucalyptus* forests and rainforests related to different 'geographic and historical origins in spite of their growing side by side' (Herbert 1932). Specifically, many authors believed that rainforests were composed of an invasive extra-Australian floristic element, an idea first put forward by the British botanist Joseph Hooker in a superb introductory essay in the *Flora of Tasmania*, published in 1860. Hooker noted that many species in rainforest on the east coast of Australia were characteristic of Indian, Polynesian, and Malayan floras. A number of workers have defined Australian rainforest according to the aggregation of biogeographically distinctive species which were assumed to have originated from beyond Australian shores (Brough *et al.* 1924; Petrie *et al.* 1929; Patton 1933; Fraser and Vickery 1937; Cromer and Pryor 1943). More recently, Webb and Tracey (1981) have argued that the Australian rainforests should be seen as being composed of indigenous, albeit ancestral, species, in which case rainforests can be seen as relictual pockets of ancestral vegetation. The complex problem of the biogeography of Australian rainforests will be considered in more detail in Chapter 12.

In an approach that mirrors the modern view of rainforests as relicts, some early workers believed rainforest could be defined as 'post-climax' communities according to the successional theory developed by the American ecologist Clements. The rainforests were thought to be relicts because they are restricted to small, localised sites (McLuckie and Petrie 1927; Davis 1936), possibly because of competitive inferiority to the dominant Australian flora. This view was ably put by Petrie *et al.* (1929), who noted that in Victoria rainforest 'frequently shows more signs of waning in competition with the virile endemics ... we are perhaps witnessing ... the closing chapters in this age-long struggle between these two great elements of the Australian flora'. The relation between rainforest and non-rainforest vegetation is considered in detail in Chapter 12.

Rainforest defined by fire susceptibility

Fire has been used to explain the restricted distribution of rainforest in Australia. Consequently a number of workers have argued that the degree of community fire sensitivity is the chief diagnostic feature in the differentiation of rainforest and non-rainforest. For example Ash (1988) defined rainforest as 'pyrophobic' and non-rainforest as pyrophytic. Cameron (1992), who sought to define rainforest in Victoria by emphasising its relative fire-sensitivity compared to non-rainforest vegetation, followed a similar strategy. His definition explicitly omits reference to many of the attributes traditionally considered essential to the definition of rainforest, such as high and/or relatively aseasonal rainfall, presence of a closed canopy, freedom from a major disturbance of external origin, absence of sclerophyll leaves, presence of characteristic special life forms (climbers and epiphytes), absence of annuals or biennials, and absence of grasses. However, Cameron excluded fire-sensitive inland communities such as *Callitris* and *Allocasuarina* forests and woodlands on structural and floristic grounds, but included transitional (ecotonal) and seral forests dominated by species such as *Eucalyptus*. The role of fire in controlling rainforest in Australia is considered in Chapters 8, 9, 10, 11 and 12.

Defining rainforest by what it is not

Given the enormous variability of rainforest as he understood it, Webb (1959) argued that Australian rainforest is best defined in negative terms. His negative definition emphasised the floristic, structural, and physiognomic differences between rainforest and the dominant Australian vegetation, with particular emphasis on sclerophylly. The definition of 'sclerophyll' when used as the primary character to separate rainforest from wet and dry sclerophyll forests has proved very problematic (Seddon 1974; Gillison 1981; Johnston and Lacey 1984; Ash 1988). Confusingly, Webb described the non-rainforest vegetation dominated by Australian taxa such as *Eucalyptus, Acacia* and Proteaceae as being 'sclerophyll' while noting that rainforest trees may also have anatomically sclerophyllous leaves. Webb (1959) explained this apparent contradiction by noting that his use of the term sclerophyll has an ecological sense and has no anatomical basis. Indeed, in a preliminary vegetation classification Webb described rainforest dominated by conifers in the family Araucariaceae as 'sclerophyll vine forest' (Herbert 1960). Subsequently, Webb (1978) expanded on this point by arguing that sclerophylly is actually a synonym for the characteristic vegetation of Australia dominated by myrtaceous, leguminous, and proteaceous species, a thought consistent with Hooker's (1860) original observation concerning the peculiar physiognomy of typical Australian vegetation. Webb (1978) suggested that non-rainforest 'sclerophyll' species are probably different anatomically and chemically to rainforest species that have hard, stiff, coriaceous leaves, but noted an absence of data to substantiate this thought. This important problem is dealt with in the following chapter.

Russell-Smith (1991) used the device of defining rainforest by what it is not in a comprehensive survey of monsoon rainforests in the northern half of the Northern Territory. In this case, the concept of rainforest excluded woody vegetation dominated by *Eucalyptus, Melaleuca, Callitris*, or mangroves. However, many of the 'rainforests' that were sampled by Russell-Smith were seasonally deciduous, some had low diversity of tree species, some were dominated by *Acacia* or included *Eucalyptus, Melaleuca* and *Callitris*, and some 'forests' were scattered patches of woody vegetation that did not form a continuous closed canopy.

Vegetation classifications that sidestep the term rainforest

Specht's (1981*a*) classification of Australian vegetation is not alone in attempting to sidestep defining problematic terms like 'rainforest' and 'sclerophyll forest' by avoiding their use. However this strategy of denial can result in new and more complex terminologies being introduced into the ecological lexicon (Gillison 1981; Johnston and Lacey 1984). Further, it is often difficult to adequately define or measure the new physiognomic terms. For example Johnston and Lacey (1984) propose the term 'hyptiophyll' to describe horizontally arranged simple or compound large leaves usually with dull green upper leaf surfaces and paler, sometimes hairy lower leaf surfaces. Such leaves are thought to be typical of many rainforest communities. Such alternative classification schemes have proven unpopular as evidenced by their limited subsequent use or citation.

Issues arising from definitional problems
Rainforest or rain forest?

The original translation of the term 'Regenwald' into English was the compound noun 'rain forest'. Although this form is routinely used elsewhere in the world, in Australia it has been contracted into one word 'rainforest'. Baur (1968) proposed this contraction because it denotes a discrete vegetation formation and avoids the 'undue emphasis on rain as the sole determining environment factor'. Webb and Tracey (1981) and Adam (1992) support the single-word spelling because it neatly flags that Australians have a unique conception of rainforest and particularly because it breaks any implied link with rainfall. Because of these precedents I will follow this admittedly confusing orographic convention throughout this book.

The problem of mixed forest

Webb's (1959) original definition of rainforest sought to emphasise differences between rainforest and non-rainforest vegetation by explicitly ignoring ecological transitions (ecotones) that contain 'sclerophyll' species such as *Acacia, Casuarina,* Myrtaceae such as *Backhousia, Eucalyptus, Syncarpia* and *Lophostemnon,* and in Tasmania epacrids and conifers. Subsequently, Webb accepted a revision of this definition to include 'transitional and seral communities with sclerophyll emergents that are of similar botanical composition to mature rainforests in which sclerophylls are absent' (Dale *et al.* 1980; Webb 1978, 1992). Webb (1978) suggested that adding the prefix 'sclerophyll' or the name of the dominant sclerophyll can classify transitional forests. The inclusion of non-rainforest trees is more consistent with the views of Fraser and Vickery (1938), who believed the occurrence of *Eucalyptus saligna* and *Syncarpia laurifolia* in their definition of subtropical rainforest demonstrated the successful fusion of two fundamentally distinct floras. Gilbert (1959) and Jackson (1968) defined mixed forests as being comprised of a tall (60–100 m) *Eucalyptus* forest above a lower (20–35 m) mature *Nothofagus* rainforest understorey. Jackson (1968) noted that in mixed *Eucalyptus–Nothofagus* forests 'the rainforest differs in no way from pure rainforest; that is if the eucalypts are ignored, the remaining vegetation is floristically and structurally rainforest'. Applying a strict interpretation of Specht's scheme to mixed forests with a low density of emergent *Eucalyptus* trees would yield the confusing definition of a 'tall woodland' of *Eucalyptus* over a 'closed forest' of *Nothofagus* (Table 2.2). To avoid such problems Walker and Hopkins (1984) suggested that mixed forests are best classified by combining both Webb's (1978) and Specht's (1981a) schemes.

Ad hoc regional definitions of Australian rainforest: Tasmanian, New South Wales and north Queensland examples

Because there has been no universally agreed definition of Australian rainforests, many workers develop their own definition of rainforest to suit local circumstances, drawing on various features of previous classifications of Australian rainforest. Three definitions of rainforest used in these schemes are described below.

Tasmania

Jarman and Brown (1983) devised a classification for Tasmanian rainforests in order to overcome a number of defects with existing systems. These defects included the exclusion of sclerophyll epacrids and coniferous species from Webb's (1978) concept of rainforest, and the axiomatic exclusion of open forests dominated by rainforest species in Specht's (1981a) scheme. They proposed a system based on vascular plant species thought able to regenerate independently of catastrophic disturbances such as fire. This definition specifically excluded mixed stands of rainforest species and *Eucalyptus*. In addition to this ecological definition, they recognised that many rainforest species are representatives of a presumed relictual flora (Nelson 1981). They used an arbitrary canopy height cut-off of 8 m to differentiate their rainforest community from alpine vegetation.

New South Wales

Floyd (1990) essentially defined rainforest by following Baur (1965) but chose to emphasise the importance of canopy cover, and the role of fire in regeneration. He defined rainforest in New South Wales as being 'a closed canopy of trees – more than 70% foliage projection of the tallest stratum – which are mainly humidity-dependent, particularly in the early stages of their life cycles, usually with more than one tree layer and with characteristic vines and epiphytes. Rainforest species are those which regenerate under shade or in natural gaps in the canopy, rather than following fire as do the genus *Eucalyptus* and most species of *Acacia*'.

North Queensland

In a survey of 'dry rainforests' in the inland regions between Cairns and Rockhampton, Fensham (1995) used the following criteria to define this vegetation type: plant communities not regularly inundated; a wet season herbaceous layer with less than 10% projective foliage cover; and a canopy in which the genera *Acacia*, *Callitris* and *Casuarina* contribute less than 50% of the total projective foliage cover. Fensham noted the strong floristic similarities of his 'dry rainforest' with rainforests in the Australian humid tropics and monsoon tropics (Russell-Smith 1991). He did not use Webb's (1978) term 'vine-thicket' to describe his communities because he found high densities of vines only in structurally degraded forests.

The fossil record and rainforest definitions

Because there is little agreement about the definition of extant rainforest and non-rainforest vegetation, it is difficult to define rainforest from fossilised leaves and pollen. Macrofossil deposits allow some inferences to be drawn about the physiognomy of localised patches of vegetation based on the spectrum of leaf sizes and types, and the relative proportions of sclerophyll and deciduous taxa (Greenwood 1994). However, differentiation of rainforest and sclerophyll communities in Webb's ecological sense is problematic for Tertiary macrofossil deposits given the currently accepted view that rainforest is the ancestral vegetation type which gave rise to the non-rainforest vegetation characteristic of modern Australia (Webb *et al.* 1986). The history of Australian rainforest and divergence of non-rainforest vegetation is discussed in Chapters 11 and 12.

Assemblages of pollen can only indicate the gross (i.e. genus and family taxonomic ranks of some but not all species) floristic composition of vegetation and carries no information concerning physiognomy beyond allowing for comparison with extant vegetation types. In practice, the identification of rainforest on the basis of pollen typically hinges on the presence of a few indicator taxa such as *Nothofagus* and *Araucaria,* the absence of some indicator taxa such as *Eucalyptus* and grasses and other herbs, and the absence of charcoal (Martin 1978).

Rainforest definition and the conflict between development and conservation

The problem of defining rainforest in Australia has not been restricted to ecology but has been debated by environmentalists, politicians and lawyers (Adam 1992). A good example was the conflict over proposed forestry development at Terania Creek in northern New South Wales in the 1970s and early 1980s. Conservationists used a broad definition of Australian rainforest that included the commercially valuable tree species *Lophostemon confertus* and *Eucalyptus pilularis* to substantiate their claim that rainforest was going to be logged. On the other hand, the Forestry Commission of New South Wales denied it was proposing to log rainforest, applying a narrower definition and arguing that *Lophostemon confertus* and *Eucalyptus pilularis* are typically components of non-rainforest vegetation. A virtually identical argument occurred over the definition of Victorian rainforest. The Victorian Government gave a commitment not to log rainforests, but excluded forests that included commercially valuable *Eucalyptus* trees (Gell and Mercer 1992). The appellation 'rainforest' has been bitterly contested because it is a familiar term to the general public that succinctly flags global threats to biological diversity (Webb 1983; Seddon 1985). Nearly all commercially valuable rainforest has been logged in Australia and most rainforest types are, comparatively speaking, well represented in conservation reserves. Environmental activists have now shifted the focus of their campaigns to stopping logging of dwindling stocks of old-growth forest regardless of whether it is defined as rainforest or not (Kirkpatrick *et al.* 1990). The definition of 'old-growth forests' has also run into semantic and legal difficulties.

Conclusion

There is no consensus as to what constitutes rainforest in Australia (Figure 2.6). This lack of consensus is attributable to a range of factors including the geographic scope of classifications (from global to regional), the purpose of the classifications, and the ecological knowledge available to the classifiers. The aim of this book is not to develop a new classification or to champion any particular existing scheme, but rather to critically evaluate ideas about the factors that

Figure 2.6
Broadleaf understorey beneath a tall Eucalyptus globulus *forest on the east coast of Tasmania.
Closed canopy vegetation beneath the tall and open canopies of* Eucalyptus *has bedevilled the
classification of rainforest in Australia. (Photograph: David Bowman.)*

control the distribution of rainforest. Given this terminological tangle, in the
following chapters I will use the original definition and description of rainforest
that was provided by the authors of research that I have selected to review. I will,
where possible, indicate the taxonomy of the dominant species of communities
that I discuss. Where appropriate, I also provide the synonyms with Webb's
(1978) classification of Australian rainforests, as given in Table 2.1. Webb's
(1978) scheme, although imperfect, is widely used and highly regarded, and it
would be wrong-headed to ignore its currency and utility. However, it should be
realised that Webb's definition of rainforest hinges on the definition of sclerophylly, a problem that is discussed in the following chapter.

3

The sclerophyll problem

In his essay on Australian phytogeography, Hooker (1860) demonstrated that the Australian vascular flora was not taxonomically unique. Instead, he concluded that the 'peculiarities of the Flora, great though they be, are found to be more apparent than real'. However, Hooker felt that many components of the Australian flora had a 'very peculiar habit or physiognomy, giving in some cases a character to the forest scenery (as *Eucalypti, Acaciae, Proteaceae, Casuarinae, Coniferae*)' with some plants having 'an anomalous or grotesque appearance' (Figure 3.1). The idea that the dominant Australian vegetation has a peculiar physiognomy including a particular prevalence of sclerophylly is central to a number of definitions used to distinguish rainforest from characteristically Australian vegetation (Francis 1951; Beadle and Costin 1952; Webb 1959, 1968, 1978). Webb differentiated non-rainforest vegetation from rainforest on the basis of sclerophylly. Confusingly, he conceded that rainforest trees species could have leaves that are anatomically sclerophyllous (Figure 3.2). He attempted to explain this contradiction by arguing that 'sclerophylly has no general anatomical justification when used in its ecological sense' although he suggested that 'sclerophyll' non-rainforest species might be physiologically and chemically distinct from rainforest plants that are also anatomically sclerophyllous. Other workers have argued that sclerophylly is essentially a physiological condition that enables non-rainforest plants to grow on infertile soils (Loveless 1961; Beadle 1966; Johnson and Briggs 1981). It is remarkable that despite the fact that there is no consensus as to the meaning of the term 'sclerophylly' it has become a central concept in Australian vegetation science. Clearly, it is tautological to base theories or classificatory systems upon a concept that cannot be defined independently. The objective of this chapter is to briefly consider the various

Figure 3.1
European botanists were astonished by the growth forms of Australian plants such as Banksia serratifolia. *A distinguishing feature of many Australian plants is their sclerophyllous foliage and distinctive floral morphologies that give rise to robust woody fruits. (Photograph: Don Franklin.)*

definitions of sclerophylly. I will then show that it is not possible to objectively differentiate rainforest from non-rainforest using any currently accepted measure of sclerophylly.

Schimper and the original use of the term sclerophyll

Not only did Schimper (1903) coin the term rainforest; he was also the first author to use the word 'sclerophyll' to denote an array of leaf characteristics possessed by a subset of plants he believed were adapted to 'physiological drought', the so-called xerophytes. According to Schimper, sclerophylly included the following features: leaf orientation (parallel or oblique to direct sunlight); leaf shape (moderately sized, typically entire, lanceolate, linear or acicular leaves, rarely compound); and leaf anatomy (heavy cuticle often with dull or bluish surface, abundant sclerenchyma, few intercellular spaces). Schimper noted that these features 'in aggregate give the leaf its characteristic, stiff, leathery consistency'. Schimper believed that vegetation in warm temperate

Figure 3.2
The foliage of the rainforest tree Syzygium australe *is anatomically sclerophyllous. (© Murray Fagg, Australian National Botanical Gardens.)*

regions with summer drought was characteristically composed of sclerophyllous plants and therefore described such vegetation as 'sclerophyllous woodland'.

The concept of sclerophylly has undergone substantial changes since Schimper's original coinage. The idea that sclerophyll leaves are necessarily xerophytic adaptations has been shown to be wrong (Beadle 1966; Seddon 1974): there is no better example than species with sclerophyll leaves that occur in, and sometimes dominate, tropical rainforests (Richards 1952; Peace and Macdonald 1981; Medina *et al.* 1990). Because sclerophyll foliage is not necessarily related to drought stress, many authors classify sclerophyllous leaves as being 'xeromorphic' rather than possessing xerophytic adaptations (Beadle 1966; Seddon 1974). Seddon (1974) reviewed the concepts of xerophytes, xeromorphs and sclerophylls and concluded that the two terms xeromorph and sclerophyll have converged to become effectively synonymous. For instance, Beadle (1966) occasionally referred to sclerophylls as 'extreme xeromorphs'. The use of the term sclerophyll free of any connotations of drought avoidance eliminates the 'terminological absurdity' (Webb 1960) associated with the term 'wet sclerophyll forest' that has wide currency in Australia.

The publication of *Die Pflanzenwelt von West Australien* by Diels in 1906 influenced how Australian ecologists came to use the term sclerophyll (Grieve 1955). Diels followed Schimper in arguing that the vegetation of southwestern Australia owed its peculiar character to the dominance of plants with sclerophyll leaves, and provided a detailed account of the foliage of numerous plants that he considered to be sclerophyllous. Building on the work of Schimper, Diels and other scholars, Grieve (1955) listed the following features of Australian sclerophylls: broad leathery leaves; microphylly; needle leaves; aphylly; winged stems; spiny stems; sunken stomata; cutinisation and lignification of the leaves; development of tannins and resinous substances; strong development of palisade mesophyll and weak development of spongy mesophyll; and presence of hairs and scales or waxy bloom on the leaf surface.

Measurement of sclerophylly

Measurement of sclerophylly includes determination of the following properties of leaves: mechanical (e.g. the ability to withstand penetration and resist fracture), chemical (e.g. moisture content, concentration of nitrogen and phosphorus and the ratio of crude fibre to crude protein), anatomical (e.g. thickness of the lamina, cuticle, palisade and epidermis and palisade/ non-palisade ratio) and physical (density, area of lamina per unit mass) (Choong *et al.* 1992; Turner 1994a). Many of these measures are strongly inter-correlated (Specht and Rundel 1990; Choong *et al.* 1992; Turner 1994a). Of the above measures, the ratio of crude fibre to crude protein has been most widely used to quantify sclerophylly, and is sometimes called the Sclerophyll Index (SI) (Specht and Rundel 1990; Turner 1994a,b). This Sclerophyll Index was developed by Loveless (1961, 1962), who suggested that crude fibre is best expressed as a proportion of leaf protoplasm (as approximated by crude protein or foliar nitrogen concentration multiplied by the constant 6.25). Foliar nitrogen was used as the denominator rather than total leaf weight because it is less variable between different species and it is independent of fibre content. Turner (1994a) showed that a leaf with a high sclerophyll index is typically thick and has a low Specific Leaf Area, a high fibre content and low concentrations of foliar nutrients.

In his Australian studies, Beadle did not use Loveless's Sclerophyll Index because he believed that quantitative definitions of sclerophylly were of less importance than physiological features that enabled plants to grow on infertile soils, particularly with extremely low levels of phosphorus. Loveless (1961, 1962) also suggested that sclerophylly has a physiological basis linked to phosphorus limitation. The exact nature of this physiological condition is not described, although Beadle (1968) speculated that the hormonal systems of mesophyllous and sclerophyllous plants might be quantitatively distinct and that limited supplies of soil phosphorus favoured the production of lignified material rather than the growth of new cells.

Beadle's failure to define response variables linked to 'sclerophylly' rendered impotent tests of his hypothesis that there was a direct relationship between different levels of plant macro-nutrients and leaf sclerophylly in non-rainforest species (Beadle 1953, 1966, 1968). For example, Beadle (1953) suggested that the addition of phosphorus and nitrate controlled sclerophylly in some plants. He illustrated this point by noting that leaves of certain sclerophyll species (particularly members of the Myrtaceae) 'can be made to increase enormously in size so that they resemble in texture the leaves of rainforest plants by the addition of these nutrients'. Subsequently, however, Beadle (1966) noted that 'sclerophylly is not necessarily correlated with leaf size'. This view was supported by anatomical studies that demonstrated that, although phosphorus and nitrogen fertilisation caused leaves to increase threefold, this growth was not accompanied by any corresponding change in leaf anatomy (Beadle 1968). There is evidence that Beadle's failure to define sclerophylly resulted in circular reasoning. For example, Beadle (1962a) conducted an experiment on the growth response of six common non-rainforest tree species on different soil types with varying levels of total soil phosphorus. Table 3.1 shows that soils with higher concentrations of total phosphorus supported increased growth of all species except *Hakea dactyloides*. Conversely, the growth of *Eucalyptus saligna* and *Eucalyptus pilularis*, species that dominate moist forests with fertile soils, were stunted on soils low in phosphorus, and Beadle concluded that these *Eucalyptus* species would be outcompeted by 'xeromorphs'. In summary, the results of Beadle's experiments proved little more than that the addition of fertiliser increased the growth rate and leaf area of some 'sclerophyll' plants.

Phosphorus toxicity (i.e. death of, or physiological damage to, plants grown under *high* concentrations of phosphorus) is evidence for a physiological adaptation of non-rainforest plants to low soil phosphorus levels. Experimental studies demonstrate that seedlings of a range of Australian non-rainforest species are sensitive to all but extremely low levels of phosphorus (Specht and Groves 1966; Grundon 1972; Groves and Keraitis 1976; Grose 1989). However, some rainforest tree species also appear to have physiologies adapted to low levels of soil phosphorus. For example, Webb (1954) found that a number of rainforest

TABLE 3.1. Mean height (cm) growth of six plant species grown in pots
with soil types with different concentrations of total phosphorus. For
each treatment there were two replicates and the total mass of soil in
each pot was 1 kg. The plants were grown in a glasshouse for five
months

Species	Soil type and total soil phosphorus concentration (ppm)			
	sandstone	laterite	shale	shale
	23	35	150	230
Eucalyptus saligna	11	48	211*	270*
Eucalyptus pilularis	11	23	105*	140*
Eucalyptus gummifera	15*	31*	74	81
Acacia suaveolens	12*	47*	131	170
Leptospermum persiciflorum	55*	55*	160	94
Hakea dactyloides	34*	32*	13	17

An asterisk denotes species which naturally occur on a given soil type.
Adapted from Beadle (1962a).

species that occur on leached acidic soils (such as the rainforest dominant
Ceratopetalum apetalum) accumulate aluminium in their foliage. He suggested
that the accumulation of foliar aluminium and associated acidic cell sap may be a
physiological adaptation to utilise the aluminium phosphate that is often the
only source of phosphorus in infertile soils.

Another possible physiological difference between rainforest and non-rainforest plants is the efficiency of nutrient withdrawal from senescent tissue. The
available data show that all non-rainforest species withdraw phosphorus from
senescent foliage, as also do many rainforest species, albeit less efficiently (Figure
3.3). Lambert and Turner (1989) found that rainforest trees withdrew phosphorus from either wood or leaves, while non-rainforest species employed both
these nutrient conservation strategies. An exception to this rule was the rainforest tree *Orites excelsa* (Figure 3.4). They suggested that the extremely efficient use
of phosphorus by *Orites excelsa* may be an adaptation to infertile soils. Beadle
(1968) also found that *Orites excelsa* was able to withdrawal phosphorus from
secondary xylem, in contrast to other rainforest species he tested and concluded
that this species was 'somewhat xeromorphic'. Such a conclusion is clear evidence of Beadle's circular reasoning concerning the dichotomisation of rainforest and sclerophyll vegetation on the basis of phosphorus.

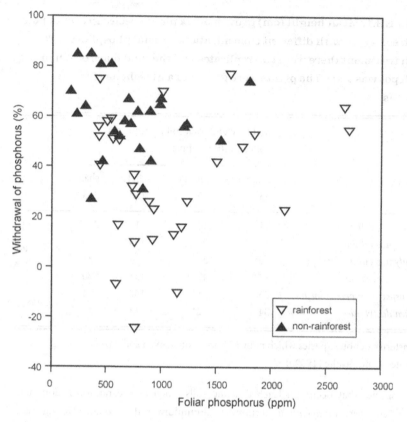

Figure 3.3

Relationship between foliar phosphorus concentrations and the % withdrawal of phosphorus upon leaf abscission for rainforest (open triangles) and non-rainforest trees (solid triangles). Withdrawal was calculated by the following formula:

$100 \times ([P$ *concentration in fresh mature leaves* $- P$ *concentration in freshly fallen leaves]/ P concentration in fresh mature leaves).*

The rainforest species were as follows: Acmena smithii*, Atherosperma moschatum*, Bombax ceiba, Brachychiton acerifolium, Caldcluvia paniculosa, Ceratopetalum apetalum*, Cryptocarya erythroxylon, Dendrocnide excelsa, Dysoxylum fraseranum, Euodia micrococca, Geissois benthamiana, Heritiera actinophylla, Heritiera trifoliolata, Neolitsea reticulata, Nothofagus cunninghamii*, Orites excelsa*, Pouteria sericea*, Sloanea woollsii*, Solanum aviculare, Solanum mauritianum*, Strychnos lucida *and* Tristaniopsis laurina*.
* = *species withdrawal rates* > 20% *and with mature foliar phosphorus concentration of* < 1000 ppm.

The non-rainforest species were as follows: Acacia aulacocarpa, Banksia serratifolia, Buchanania obovata, Erythrophleum chlorostachys, Eucalyptus baxteri, E. confertiflora, E. grandis, E. globoidea, E. gummifera, E. miniata, E. nitida, E. obliqua, E. pilularis, E. polyanthemos, E. regnans, E. sieberi, E. tetrodonta, Lophostemon confertus, Olearia argophylla, Persoonia laevis, Planchonia careya, Pomaderris aspera *and* Terminalia ferdinandiana.

Derived from: Ashton and Frankenberg (1976), Ashton (1975, 1976a), Beadle (1968), Bowman et al. *(1986), Fensham and Bowman (1995), Lambert and Turner (1983, 1989), Melick (1990b) and Turner and Kelly (1981).*

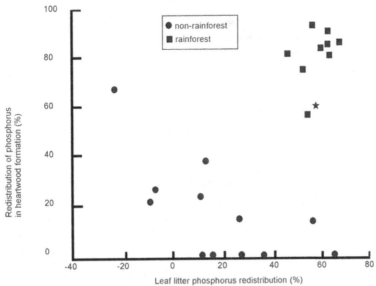

Figure 3.4

Relationship between withdrawal of phosphorus from senescent leaves and heartwood in a range of subtropical rainforest and non-rainforest tree species. The sclerophyll tree species are as follows: Casuarina torulosa, Eucalyptus dives, Eucalyptus grandis, Eucalyptus maculosa, Eucalyptus obliqua, Eucalyptus pilularis, Eucalyptus rossi, Eucalyptus rubida *and* Lophostemon confertus. *The rainforest species are as follows:* Caldcluvia paniculosa, Ceratopetalum apetalum, Cryptocarya erythroxylon, Dendrocnide excelsa, Dysoxylum fraseranum, Euodia micrococca, Geissois benthamiana, Heritiera actinophylla, Heritiera trifoliolata, Neolitsea reticulata, Orites excelsa *(marked with *),* Sloanea woollsii, Solanum aviculare *and* Solanum mauritianum. *Adapted from Lambert and Turner (1989).*

Case studies in the determination of sclerophylly in Australian woody plants

Given the importance of the term 'sclerophyll' in defining Australian forest types, it is remarkable how few studies have been made of the physical, anatomical, and chemical characteristics of the leaves of Australian trees (Webb 1978). Below I summarise results of various studies on the leaf characteristics of rainforest and non-rainforest trees.

Leaf texture across a humid rainforest disturbance gradient

Using a three-point scale of leaf texture (mesophyll, intermediate or sclerophyll), Westman (1990) demonstrated high levels of sclerophylly throughout a gradient of increasing time since disturbance in humid tropical rainforest on the Atherton Tablelands. Surprisingly, the recently disturbed stand was found to be less sclerophyll and had smaller-leaved plants than mature rainforest (Figures 3.5 and 3.6).

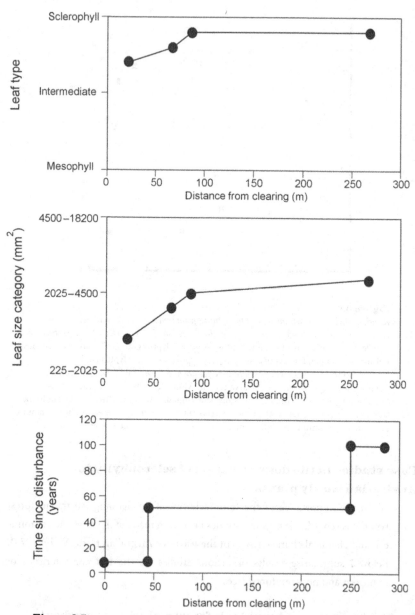

Figure 3.5
Average values of leaf type and leaf size, each calculated on a three-point scale, at four sites across a disturbance gradient at Lake Eacham on the Atherton Tablelands. Adapted from Westman (1990).

Figure 3.6
The mesophyllous giant stinging tree Dendrocnide excelsa *is characteristic of disturbed areas of tropical and subtropical rainforest. (Photograph: Don Franklin.)*

Leaf anatomy and a rainforest–*Eucalyptus* boundary

McLuckie and Petrie (1927) compared the leaf anatomy of dominant trees (*Atherosperma moschatum, Doryphora sassafras, Ceratopetalum apetalum*) in rainforest at Mount Wilson in the Blue Mountains, New South Wales to those of a surrounding *Eucalyptus* forest. They demonstrated that the rainforest trees had mesophytic leaves with thin cuticles, limited thickening of cell walls and well-developed spongy mesophyll (Figure 3.7). In the surrounding *Eucalyptus* forest, the leaf anatomy of characteristic plants was found to be highly variable (Figure 3.7). For example, *Telopea speciosissima* had thick cuticles, thickened cell walls in the palisade layer and sunken stomata which are restricted to the lower leaf surface. The upper leaf surface of *Banksia serrata* leaves had very thick smooth cuticles, while the lower leaf surface contained cavities with loose twisted hairs beneath which occurred raised stomata. Girders of sclerenchyma that

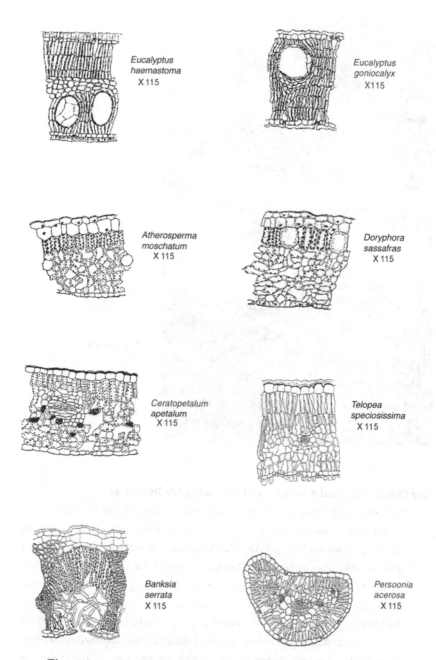

Figure 3.7
Anatomical sections of leaves of selected rainforest and non-rainforest plants at Mt Wilson, NSW. Adapted from McLuckie and Petrie (1927).

surrounded the vascular bundles strengthened the leaves. In contrast, *Persoonia acerosa* had slightly succulent leaves with occasional sclerides. McLuckie and Petrie also found significant anatomical differences (number of stomata, number of oil glands, thickness of cuticle and density of palisade tissue) in the foliage of *Eucalyptus* species grown on different substrates. *Eucalyptus haemastoma*, which occurred on infertile sandstone soils, had an 'enormously thickened cuticle' and a central aqueous zone between the dense layers of upper and lower palisade. In contrast, *Eucalyptus goniocalyx*, which occurred on fertile basalt soils, had loosely arranged palisade tissue which was not separated by a central aqueous layer, and a thinner cuticle and outer epidermal walls and fewer oil glands than *Eucalyptus haemastoma* (Figure 3.7). These data demonstrated that some *Eucalyptus* species were anatomically mesophytic, a conclusion also supported by anatomical sections of adult *Eucalyptus fastigata* and *Eucalyptus regnans* leaves (Cameron 1970; Ashton and Turner 1979) (Figure 3.8). Both *Eucalyptus regnans* and *E. fastigata* occur in high rainfall climates that were thought by Schimper (1903) to support temperate rainforest.

A problem with characterising species as sclerophyll or mesophyll is that there is enormous variation in the leaf anatomy within the one tree species. For example, Myers *et al.* (1987) demonstrated that leaf anatomy of the tropical rainforest tree *Castanospermum australe* varied significantly with canopy position. Understorey leaves were far more mesophytic (i.e. had higher specific area, larger epidermal cells and thinner leaves with large intercellular spaces) than canopy or open grown leaves (Table 3.2). The age of plants and of individual leaves also affects the degree of sclerophylly. For example, Ashton and Turner (1979) showed that leaf anatomy in *E. regnans* varied substantially with the age of the tree (Figure 3.8). Lowman and Box (1983) demonstrated that leaves of some rainforest trees became more resistant to penetration once the leaves had matured fully more than a year after emerging (Figure 3.9). In the case of the sclerophyll species *Hakea dactyloides*, Hamilton (1914) noted anatomical changes with leaf age, with almost half the cells in a mature leaf being composed of sclerenchyma while immature leaves contained few sclerids.

Leaf secondary compounds and herbivores across rainforest–*Eucalyptus* boundaries in central NSW

The secondary compound data of Macauley and Fox (1980) and Lowman and Box (1983) shed some light on the proposition that there are biochemical differences in the foliage of rainforest and *Eucalyptus* species (Table 3.3). Healthy mature leaves from rainforest species had a lower concentration of total

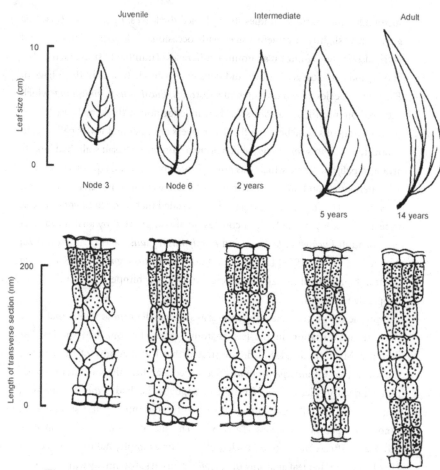

Figure 3.8

Anatomical sections of Eucalyptus regnans *leaves associated with ontogenic development. With increasing tree age leaves change from having contrasting adaxial and abaxial anatomies associated with a more horizonal orientation to an isobilateral anatomy associated with vertically arranged foliage. At no stage are the leaves truly sclerophyllous (adapted from Ashton and Turner 1979).*

phenolics and a higher proportion of condensed tannins than the *Eucalyptus* species. Such comparisons are complicated by the fact that there are marked effects of light (full sun or shade), season and leaf age on the concentrations of phenolic substances in individual leaves (Macauley and Fox 1980; Lowman and Box 1983). Without more data systematically collected from a wide range of rainforest and non-rainforest species it is impossible to know if the patterns in Table 3.3 are representative of real differences. Comparison of biochemical characteristics of foliage from taxa, such as *Syzygium*, that are phylogenetically

TABLE 3.2. Mean (standard error) leaf structure characteristics of the north Queensland rainforest tree *Castanospermum australe* grown in three contrasting microclimates

Microclimate	Surface area of epidermal cells (μm²)	Percentage of cell wall composed of epidermal cells	Specific leaf area (m² kg⁻¹)
Open-grown	23.8 (0.8)	23.9 (0.7)	9.3 (0.2)
Canopy	16.3 (0.3)	22.7 (0.3)	9.6 (0.2)
Understorey	45.5 (1.4)	10.4 (0.3)	24.1 (0.2)
Least significant difference $P < 0.001$	4.7	1.2	0.8

Adapted from Myers *et al.* (1987).

TABLE 3.3. Phenolic and condensed tannins composition of 1–2-month-old fully expanded shade leaves from seven species of *Eucalyptus* (Macauley and Fox 1980) and 1- month to 1-year-old fully expanded leaves from five rainforest species (Lowman and Box 1983). The percentage of the total phenols that were condensed tannins is also shown

Species	Total phenols (% leaf dry weight)	Condensed tannins (% leaf dry weight)	% condensed tannins
Doryphora sassafras	2.6	1.3	50
Ceratopetalum apetalum	3.3	1	30
Toona australis	13.3	10.4	78
Nothofagus moorei	11	8.2	75
Dendrocnide excelsa	1.5	1	66
Eucalyptus blakelyi	28	6	21
Eucalyptus mannifera	30.5	13	43
Eucalyptus delegatensis	23	13	57
Eucalyptus fraxinoides	20	5	25
Eucalyptus pauciflora	16	11.5	72
Eucalyptus macrorhyncha	29	6.5	22
Eucalyptus camphora	26.5	6	23

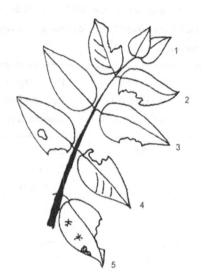

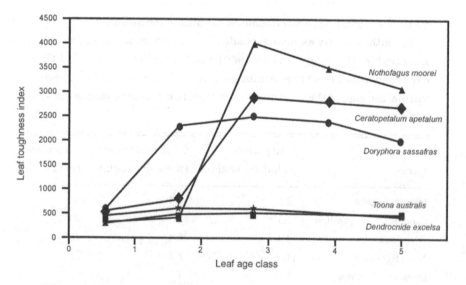

Figure 3.9

Changes in leaf toughness with increasing leaf age for five subtropical rainforest species (A Nothofagus moorei, B Ceratopetalum apetalum, C Doryphora sassafras, D Toona australis, E Dendrocnide excelsa). The leaf age categories are as follows: 1: young leaf < 2 weeks old; 2: youthful leaf 2–4 weeks old; 3: mature leaf 4–52 weeks old; 4: old leaf > 52 weeks old; 5: senescent leaf. Adapted from Lowman and Box (1983).

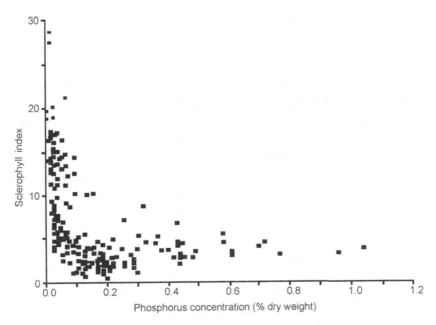

Figure 3.10
Relationship between foliar phosphorus and the Sclerophyll Index for plant species from a range of habitats throughout the world. Adapted from Turner (1994b).

related and that occur in rainforest or non-rainforest environments may provide insights into evolutionary patterns. Future comparative studies should also include studies of essential oils. The above studies, however, do not support Janzen's (1974) hypothesis that foliage rich in secondary compounds is necessarily a defence against herbivory. Lowman and Box (1983) found less insect herbivore damage to the foliage of the rainforest species they studied than had been reported for the foliage of *Eucalyptus*. These authors conclude that 'other factors such as leaf nutritive qualities, abundance and seasonality of herbivores, or herbivore predators and parasites in rainforests versus sclerophyll woodlands may be involved'. These data also indicate a long evolutionary relationship between *Eucalyptus* and its herbivores.

Concentrations of phosphorus in rainforest and *Eucalyptus* foliage

In contrast to the limited inquiry into the anatomy and biochemistry of Australian forest trees, more is known about their concentrations of macro-nutrients, particularly phosphorus. Given that the Sclerophyll Index dramatically increases when foliar phosphorus is below 1000 ppm (Turner 1994b) (Figure 3.10), it is possible to use foliar phosphorus as a proxy for the Sclerophyll Index. The following case studies are reported.

Foliar phosphorus variation within a subtropical rainforest, and differences between a wet and dry sclerophyll forest

In a study of a stand of subtropical rainforest, Lambert and Turner (1986) found that the mean foliage concentration of phosphorus was 1119 ppm. There were, however, differences between large trees (mean 945 ppm), small trees (mean 1180 ppm), large shrubs (mean 1450 ppm) and small shrubs (mean 902 ppm). According to the 1000 ppm phosphorus threshold for sclerophylly suggested by Turner (1994*b*), the large tree and small shrub layer could be described as close to sclerophyllous.

Figure 3.11 (opposite)

Histogram showing the phosphorus concentration in the foliage of Australian vascular plant species. The 1000 ppm level thought to discriminate sclerophyll from non-sclerophyll foliage is indicated with the arrow.

The Eucalyptus *species (n = 125) are:* E. agglomerata, E. angophoroides, E. badjensis, E. baxteri, E. behriana, E. bosistoana, E. botryoides, E. bridgesiana, E. calophylla, E. confertiflora, E. consideniana, E. cypellocarpa, E. dalrympleana, E. diversifolia, E. dives, E. elata, E. fastigata, E. foecunda, E. fraxinoides, E. globoidea, E. gummifera, E. haemastoma, E. incrassata, E. leucoxylon, E. longifolia, E. macrorhyncha, E. maculata, E. maidenii, E. marginata, E. microcarpa, E. miniata, E. muellerana, E. nitens, E. nitida, E. obliqua, E. ovata, E. pauciflora, E. pilularis, E. piperita, E. planchoniana, E. polyanthemos, E. radiata, E. regnans, E. rubida, E. saligna, E. sideroxylon, E. sieberi, E. signata, E. smithii, E. socialis, E. stellulata, E. tetrodonta, E. umbra *and* E. viminalis.

The non-rainforest species (n = 104) are in the following genera: Acacia, Actinotus, Allocasuarina, Angophora, Baeckea, Banksia, Buchanania, Calytrix, Casuarina, Erythrophleum, Grevillea, Hakea, Hibbertia, Lambertia, Leptospermum, Leucopogon, Lomatia, Lophostemon, Macrozamia, Melaleuca, Monotoca, Olearia, Persoonia, Petrophile, Phyllota, Pittosporum, Planchonia, Platylobium, Pomaderris, Pultenaea, Rhagodia, Syncarpia, Terminalia *and* Xanthorrhoea.

The rainforest species (n = 192) *are in the following genera:* Acmena, Akania, Alangium, Anthocarpa, Archontophoenix, Argyrodendron, Atherosperma, Baloghia, Bombax, Brachychiton, Breynia, Caldcluvia, Caustis, Ceratopetalum, Cinnamomum, Claoxylon, Cleistanthus, Cryptocarya, Daphnandra, Dendrocnide, Diospyros, Diploglottis, Doryphora, Duboisia, Dysoxylum, Ehretia, Elaeocarpus, Eucryphia, Euodia, Eupomatia, Ficus, Flindersia, Geissois, Guilfoylia, Helichrysum, Litsea, Melicope, Neolitsea, Nothofagus, Olearia, Omolanthus, Orites, Pennantia, Phytolacca, Pittosporum, Planchonella, Polyosma, Polyscias, Pouteria, Psychotria, Quintinia, Randia, Rapanea, Rhodamnia, Sarcopteryx, Sloanea, Solanum, Stenocarpus, Strychnos, Synoum, Syzygium, Tasmannia, Toona, Trochocarpa, Wilkiea *and* Zanthoxylum.

Derived from: Ashton and Frankenberg (1976), Ashton (1975, 1976a), Attiwill (1968), Baur (1957), Beadle (1954, 1968), Bowman et al. (1986), Braithwaite et al. (1983), Feller (1980), Fensham and Bowman (1995), Hocking (1982, 1986), Lambert and Turner (1983, 1986, 1987), Lambert et al. (1983), Langkamp et al. (1982), Langkamp and Dalling (1982), McColl and Humpreys (1967), Melick (1990b), Neilsen and Palzer (1977), O'Connell et al. (1978), Specht and Rundel (1990), Turner and Kelly (1981) and Westman and Rogers (1977).

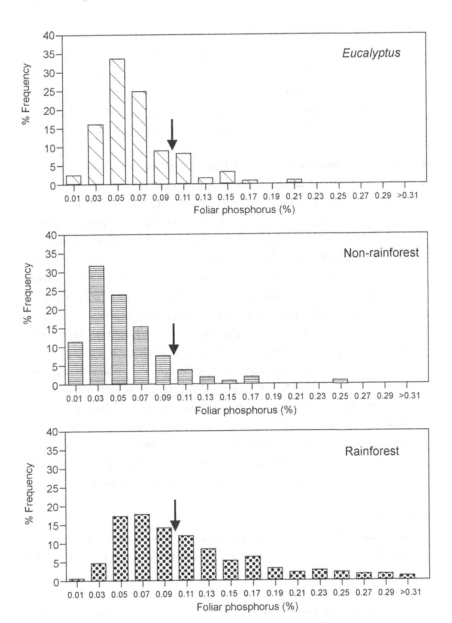

Lambert and Turner (1987) demonstrated that there was a significant difference between the foliar phosphorus concentrations in the 10 wet sclerophyll species (mean 990, range 390–2440 ppm) and 15 dry sclerophyll forest species (mean 420, range 290–630 ppm). However, there was considerable overlap between the phosphorus concentration in the foliage of wet sclerophyll and rainforest species. This point is considered in more detail below.

Distribution of published foliar phosphorus concentrations for rainforest, *Eucalyptus* and non-rainforest species

Figure 3.11 shows the frequency distribution of phosphorus values in the foliage of rainforest, *Eucalyptus* and other non-rainforest tree species compiled from a range of publications. Statistical analysis of these data with non-parametric Wilcoxon tests show that the rainforest species have significantly ($P < 0.001$) higher concentrations of foliar phosphorus than the *Eucalyptus* and non-rainforest groups of species. Importantly however, 54% of rainforest species had foliar phosphorus levels below 1000 ppm and 22% had foliar phosphorus levels below 600 ppm. Such low levels of foliar phosphorus are consistent with the observation that many rainforest plants have sclerophyllous foliage.

A feature of the data used in constructing the figure is considerable amongst-species variation of the foliar phosphorus concentration for both rainforest and *Eucalyptus* species. The source of the variation is unclear. Lambert *et al.* (1983) showed that, of the 13 species they measured in a subtropical rainforest, five had foliage phosphorus concentrations that were less than 60% (i.e. 820 ppm) of the mean surface soil phosphorus concentration (1365 ppm), suggesting that foliar phosphorus is not directly determined by local nutrient conditions.

Conclusion

There has not been a comprehensive analysis of how the anatomical, chemical and physical characteristics of foliage change across rainforest boundaries. The sparse available data, however, show that many rainforest species are 'sclerophyll' in the sense of having hard leaves and low phosphorus concentration in their foliage. There is no better example of this than the *Allosyncarpia ternata* rainforests in the Northern Territory, where the canopy is dominated exclusively by a sclerophyll, myrtaceaeous tree (Fordyce *et al.* 1995). Conversely, some *Eucalyptus* species have anatomical features and concentrations of foliar phosphorus typical of mesophyll leaves. Therefore, the generalisation that rainforest can be differentiated from adjacent non-rainforest communities on the basis of sclerophylly cannot be supported without much more research. Clearly, such research depends upon a measurable definition of sclerophylly such as Loveless's Sclerophyll Index. I concede that it is possible that the *degree* of sclerophylly is greater in non-rainforest vegetation, particularly across abrupt boundaries associated with marked differences in edaphic moisture supply. Such apparent differences in foliage characteristics across rainforest boundaries may explain why most ecologists have passively accepted the dichotomisation of rainforest and non-rainforest vegetation on the basis of sclerophylly despite the absence of any compelling evidence to bolster this view. I suspect, however, that the underlining reason why ecologists have embraced the idea that suppression of sclerophylly can be used to define rainforest harks back to Hooker's idea of Australian vegetation having a peculiar morphology and physiognomy. Fossils of sclerophyll species are known throughout the Tertiary, often occurring in assemblages interpreted as representing 'rainforest' vegetation, confounding neat dichotomisation of non-rainforest and rainforest vegetation in terms of sclerophylly. Emphasis on contrasting foliage characteristics of rainforest and non-rainforest plants has deflected attention from the ubiquity of vegetation dominated by *Eucalyptus* and *Acacia* in Australia. Why should 'sclerophyll' foliage be so widespread throughout Australia, occurring as it does from arid to humid climates? Adamson and Osborn (1924) perceptively noted this puzzle. While accepting that the dominant Australian genera *Eucalyptus* and *Acacia* have leaves that closely approach 'the so-called typical sclerophyll', they noted that 'this type of leaf, in Australia, is possessed by trees occurring in such a variety of climate and habitat that it has no indicator value'. Beadle developed a theory that sclerophylly is an adaptation to infertile soils and that rainforests are axiomatically incapable of growing on infertile substrates. This theory will be considered in the following chapter.

4

The edaphic theory I. The control of rainforest by soil phosphorus

In an insightful paper, Andrews (1916) suggested that the dominant Australian flora was originally derived from rainforest and had undergone extreme morphological modifications in order to cope with infertile soils. This view was subsequently elaborated by the ecologist Beadle (1953, 1954, 1962a,b, 1966, 1968), who argued that the restricted distribution of Australian rainforests is caused by the widespread exhaustion of soil phosphorus through geological time. Beadle (1966) took this view to the extreme by stating that 'the absence of rainforest and the paucity of rainforest genera in south-west Western Australia are accounted for by low [soil] fertility levels rather than by a decimation of rainforest genera by a past dry climate'. As discussed in the previous chapter, Beadle argued that the non-rainforest 'sclerophyll' vegetation owed its dominance to physiological characteristics that enabled it to survive on infertile soils, particularly those with extremely low levels of phosphorus.

Beadle (1953, 1954, 1962a, 1966, 1968) supported his hypothesis by examination of correlations and some experimentation. Although his hypothesis has received general support (Ashton 1976a; Attiwill *et al.* 1978; Webb 1969) and only limited criticism (Coaldrake and Haydock 1958; Le Brocque and Buckney 1994), there have been few tests of the hypothesis (Coaldrake and Haydock 1958; Richards 1968; Adam *et al.* 1989; Le Brocque and Buckney 1994). The purpose of this chapter is to demonstrate that available data do not support the generalisation that rainforest is restricted to soils with high soil phosphorus levels.

Correlation of the richness of rainforest taxa with soil phosphorus

A prediction of Beadle's hypothesis is that the diversity of rainforest taxa is positively correlated to soil phosphorus irrespective of rainfall while the converse is true for sclerophyll taxa. He acknowledged that this relationship would not hold for non-rainforest communities in cold climates, halophytic communities, and climatically dry areas with heavy textured soils. Using data collated from various sources, he showed that when total soil phosphorus levels exceeded a threshold of 200 ppm rainforest genera outnumber non-rainforest genera (Figure 4.1). He defined species that occur in non-rainforest communities such as *Eucalyptus* forests or desert scrubs as belonging to a 'rainforest' genus if most of their congeners occurred in rainforest. Although such species were morphologically different from their rainforest congeners he noted that they typically retained broad, sometimes compound, mesomorphic leaves. For example, he suggested that in arid central Australia there are about 100 species from 35 'rainforest' genera (Table 4.1).

Recent research in the Sydney region provides some support for the concept that there is a relationship between taxonomic richness of rainforest and non-rainforest vegetation and total soil phosphorus concentrations. In a study of coastal sea cliffs and headlands around Sydney, New South Wales, Adam *et al.* (1989) demonstrated a statistically significant inverse geometric relationship between non-rainforest species richness and total soil phosphorus (Figure 4.2). This curve had an inflection at about 200 ppm of phosphorus, conforming to Beadle's threshold for the diversity of sclerophyll genera (Figure 4.1). Also working in the Sydney area, Clements (1983) showed that the foliage cover of mesic species characteristic of rainforest and wet sclerophyll vegetation was positively related to total soil phosphorus concentrations. Conversely, she found the cover of xeric species characteristic of dry sclerophyll vegetation was negatively related to total soil phosphorus (Figure 4.3). Although Clement's data provided no information concerning species richness, the pattern is broadly consistent with Beadle's hypothesis although the changeover from xeric to mesic dominance occurred at around 100 ppm of total soil phosphorus. However, a feature of Figure 4.3 that is inconsistent with the Beadle hypothesis is that some sites with less than 100 ppm of total soil phosphorus have a relatively high mesic canopy cover. Multivariate analyses showed that in addition to soil phosphorus, topographic position and moisture supply also influenced mesic and xeric canopy cover (Figure 4.4).

However, data from 144 rainforest boundaries in northwestern Australia

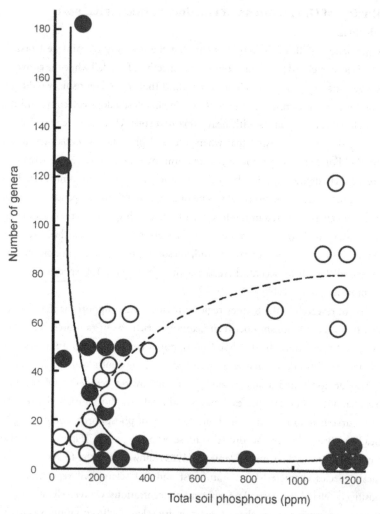

Figure 4.1

Relationship between the number of rainforest and non-rainforest genera and soil phosphorus concentrations. The dashed and solid lines were fitted by hand to depict the relationship between soil phosphorus and rainforest (open circles) and non-rainforest (closed circles) genera respectively. The data were collated from a variety of sources by Beadle (1966).

(Bowman *et al.* 1991*a*; Bowman 1992*a*) showed no correlation of the number of tree species in savanna vegetation with the concentration of total soil phosphorus. Most savannas had soils with less than 500 ppm of phosphorus, a pattern consistent with Beadle's hypothesis (Figure 4.5). In the case of monsoon rainforest vegetation, however, there is no clear relationship between the richness of tree species and total soil phosphorus levels and, like the savannas, most soils had

Table 4.1. Number of species in rainforest genera in southwestern and
central Australia

	Southwestern Australia	Central Australia
Abutilon	1	5
Alstonia	0	1
Aphanopetalum	1	0
Atalaya	0	1
Brachychiton	0	2
Canthium	0	4
Capparis	1	4
Carissa	0	1
Cassia	1	18
Cissus	1	0
Citrus	0	1
Clematis	1	0
Clerodendrum	0	1
Codonocarpus	0	1
Commersonia	0	2
Corchorus	0	2
Cynanchum	0	1
Dioscorea	1	0
Duboisia	0	1
Ehretia	0	1
Ficus	0	1
Flindersia	0	1
Geijera	1	1
Heterodendrum	0	2
Hibiscus	6	13
Jasminum	1	2
Keraudrenia	0	2
Livistona	0	1
Marsdenia	0	1
Owenia	0	2
Pandorea	0	1
Parsonsia	0	1
Pentatropis	0	3
Phyllanthus	3	5
Pittosporum	0	1
Rulingia	8	3
Solanum	3	14
Trema	0	1
Ventilago	0	1
Total of number of genera	13	35
Total number of species	29	102

Adapted from Beadle (1966).

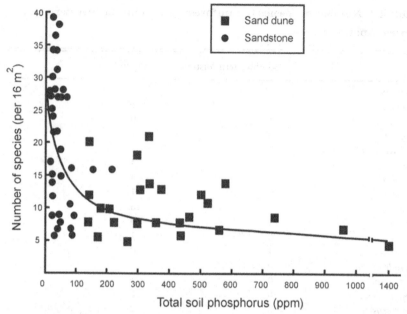

Figure 4.2
Relationship between the number of vascular plant species (both native and non-native) per 16 m² and soil phosphorus levels for coastal non-rainforest vegetation in the Sydney region on sand dune and sandstone substrates. The solid line is a regression showing the relationship between the number of native species and total concentrations of soil phosphorus. Adapted from Adam et al. *(1989).*

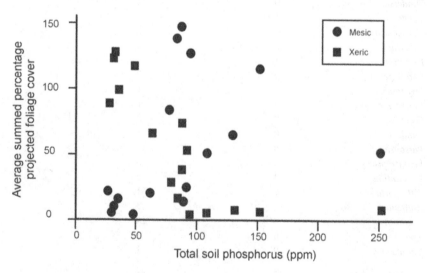

Figure 4.3
Relationships of total soil phosphorus and mean canopy cover of xeric and mesic vegetation. Adapted from Clements (1983).

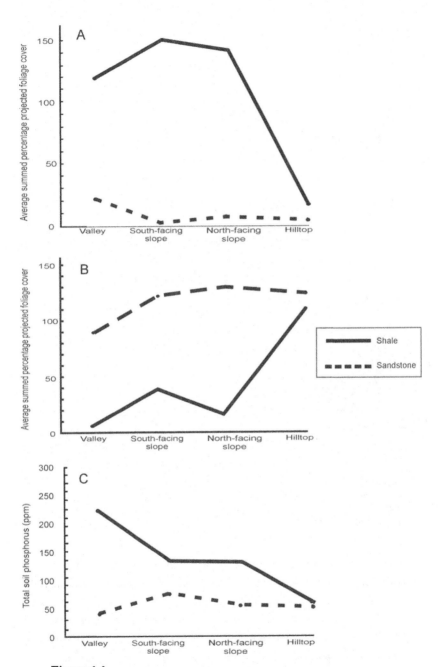

Figure 4.4

Relationship of topographic position and aspect to mean canopy cover of (A) 'mesic' and (B) 'xeric' (including rainforest species) vegetation sampled in 250 m² quadrats and total concentration of soil phosphorus (C) for forests in the Sydney region. Adapted from Clements (1983).

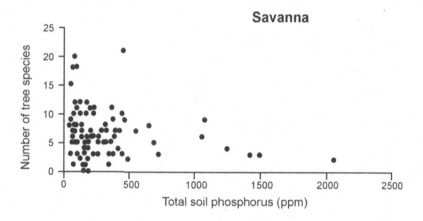

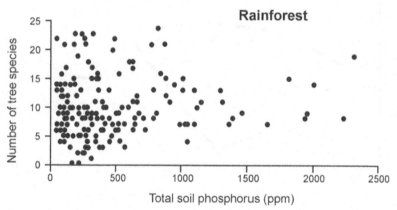

Figure 4.5
Relationship between the number of tree species (> 1.3 m in height) in 10 × 20 m quadrats and total soil phosphorus concentrations in northwest Australian savanna and monsoon rainforest. Derived from Bowman (1992a).

total soil phosphorus levels of below 500 ppm. Indeed, some of the most floristically diverse monsoon forests studied occurred on soils with less than 200 ppm. This finding is in concordance with tree species richness patterns in Tasmania where the least species-rich rainforests occur on the most fertile soils (Read 1995). However, this is not the case in the subtropics, where fertile soils are known to support species-rich vegetation while infertile soils are often dominated by a single species, *Ceratopetalum apetalum* (Turner and Kelly 1981) (Figure 4.6). The available evidence does not support Beadle's generalisation that the diversity of Australian rainforests species is positively correlated with soil

Figure 4.6
Profile of a monodominant Ceratopetalum apetalum *rainforest on the Gibraltar Range, New South Wales. This rainforest type is characteristic of soils with low levels of phosphorus. (© Murray Fagg, Australian National Botanical Gardens.)*

phosphorus concentrations. Rather, this view appears to hold only in specific environments such as subtropical eastern Australia where Beadle formulated his hypothesis.

Restriction of rainforest to soils with high soil phosphorus

Beadle's conjecture that soil fertility controls the distribution of various vegetation types in Australia has received limited critical analysis. In a study specifically conducted to test the Beadle hypothesis, Coaldrake and Haydock (1958) concluded that soil phosphorus concentrations did not determine the distribution of various non-rainforest vegetation types such as heath and *Eucalyptus, Angophora,* and *Melaleuca* dominated communities in an area of southeast Queensland. Beadle (1962*a*) dismissed their rejection of his hypothesis, arguing that their data actually supported his theory that soil '*phosphate levels determine the kind of vegetation*' (original emphasis). The source of the disagreement centred on spatial scale and the classification of vegetation. Coaldrake and Haydock (1958) tested the Beadle hypothesis by comparing a range of different non-rainforest communities within one area, while Beadle (1962*a*) argued that all these communities belonged to a single vegetation type.

There are a number of published studies that have compared the phosphorus concentration of soils across rainforest boundaries. Table 4.2 shows that there are as many published examples of rainforest vegetation occurring on soils with equivalent or significantly lower concentrations of phosphorus compared with surrounding non-rainforest vegetation as there are examples of the converse predicted by Beadle's hypothesis. A feature of these data is the enormous variability in the phosphorus concentration within and between rainforest and non-rainforest vegetation. Although analytical methodologies varied between the different studies reported in Table 4.2, there is no doubt the variation is real. For example, the variation of total soil phosphorus (157–2487 ppm) in different types of rainforest in New South Wales reported by Baur (1957) is comparable with the variation of soil phosphorus presented in Table 4.2. Given such variation of rainforest soil phosphorus there is often considerable overlap with non-rainforest soil phosphorus: four examples of this are considered below.

i. *Ceratopetalum* rainforest in the Sydney region.

Beadle (1962*a*) showed that soils in *Ceratopetalum* rainforest on Hawkesbury sandstone had phosphorus concentrations greater than non-rainforest soils on the same rock type (ranges of 65–163 *vs* 23–53 ppm respectively). However, phosphorus concentrations in Hawkesbury sandstone soils supporting *Ceratopetalum* rainforest overlapped with the phosphorus levels he reported for wet sclerophyll forest soils developed from Narrabeen shales and Narrabeen sandstones (102–230 ppm). The large overlap in soil phosphorus concentrations between soils of different vegetation types and parent materials in the Sydney region has been confirmed by Le Brocque and Buckney (1994). They demonstrated that phosphorus concentrations of *Ceratopetalum apetalum* rainforest surface soils on shale substrates were not significantly different from phosphorus concentrations in the surface soils of many different vegetation types on sandstone and shale substrates, the only exceptions being dry *Eucalyptus* communities.

ii. Subtropical rainforest.

Turner and Kelly (1981) found that mean total phosphorus concentrations in subtropical rainforest soils (1340 ppm) were similar to adjacent moist *Eucalyptus grandis* forest soils (1095 ppm) but much higher than drier *Eucalyptus pilularis* forest soils (195 ppm).

iii. Semi-evergreen vine thicket (SEVT) rainforest.

Working at the inland limit of rainforest vegetation in southeastern Queensland, Ahern and Macnish (1983) found that rainforest soils had higher concentrations of phosphorus than surrounding non-rainforest soils when these comparisons are broken down into four different parent materials. However, when the four parent materials were considered together there was substantial overlap between phosphorus levels of rainforest and non-rainforest soils (Table 4.2). A similar pattern was found by Bowman (1992*a*) in northwestern Australia. He demonstrated that dry monsoon rainforest soils had significantly higher concentrations of phosphorus than surrounding savanna (means of 890 and 360 ppm respectively) but that these mean values had extremely high coefficients of variation (279% and 111% respectively).

iv. *Nothofagus* rainforest.

Ashton (1976*a*) reported that the concentration of total phosphorus in the surface soils of a *Nothofagus cunninghamii* rainforest (range 150–360 ppm) overlapped with that of dry (north-facing) *Eucalyptus* forest soils (range 70–200 ppm) and wet south-facing *Eucalyptus* forest soils (range 240–310 ppm) in a granite drainage basin in central Victoria. Ashton concluded that *Nothofagus cunninghamii* rainforest is not limited by soil phosphorus. This view was corroborated by Ellis and Graley (1987), who found that *Nothofagus cunninghamii* rainforest in northeastern Tasmania occurred on soils with equivalent phosphorus concentrations to surrounding *Acacia*, *Eucalyptus* and *Leptospermum* forests, but had significantly lower concentration of phosphorus than adjacent grasslands.

In summary, the published data do not support the generalisation that Australian rainforests are limited to soils with high phosphorus concentration. Indeed, Le Brocque and Buckney's (1994) research in the same environment where Beadle formulated his hypothesis led them to conclude that soil phosphorus concentrations 'explain little of the major vegetation gradients within the Hawkesbury Sandstone and Narrabeen soils'.

Growth experiments, phosphorus and rainforest and non-rainforest soil

Growth experiments that Beadle and other authors have conducted do not clearly support the soil phosphorus hypothesis. For example, working in the Sydney region, Beadle (1953, 1954) found that in order to get an equivalent

TABLE 4.2. Summary of total soil phosphorus concentrations in the surface soil of various types of rainforest, and comparisons ($P < 0.05$) with those of adjacent non-rainforest vegetation

Rainforest type	Total soil phosphorus (ppm)		Statistical significance of comparisons with non-rainforest			Reference
	Mean	Coefficient of variation (%)	> non-rainforest	N.S. non-rainforest	< non-rainforest	
Podocarpus lawrencei	1131	49	X			P. Barker (1991)
Semi-evergreen vine thicket on Walloon coal measures	1350	N.A.	X			Ahern and Macnish (1983)
Semi-evergreen vine thicket on black earths	2000	N.A.	X			Ahern and Macnish (1983)
Semi-evergreen vine thicket on euchrozems	2200	N.A.	X			Ahern and Macnish (1983)
Semi-evergreen vine thicket on lithosols	3200	N.A.	X			Ahern and Macnish (1983)
Dry monsoon forest	890	279	X			Bowman (1992a)
Ceratopetalum apetalum	110	115	X	X		Le Brocque and Buckney (1994)

Nothofagus cunninghamii	240	34	X	X		Ashton (1976a)
Argyrodendron trifoliatum subtropical rainforest	1340	N.A.	X	X		Turner and Kelly (1981)
Allosyncarpia ternata	58	24		X		Bowman (1991a)
Elaeocarpus–Atherosperma	210	106		X		P. Barker (1991)
Wet monsoon forest	270	118		X		Bowman (1992a)
Dry monsoon forest	590	11		X		Bowman and Fensham (1991)
Ceratopetalum apetalum	340	N.A.		X	X	Turner and Kelly (1981)
Archontophoenix cunninghamiana palm forest	370	N.A.		X	X	Turner and Kelly (1981)
Nothofagus cunninghamii	450	N.A.		X	X	Ellis and Graley (1987)

N.A.: not available

TABLE 4.3. Mean biomass of three flax plants grown in 1 kg of different soil types

Four treatments were applied to each soil type: control, addition of 250 mg of NO_3, addition of 200 mg PO_4, and a combination of both phosphorus and nitrogen treatments. For each combination of treatments there were two replicates. The flax plants were grown in a greenhouse for six weeks.

Vegetation type	Locality and geology	Control	N	P	N + P
Ceratopetalum apetalum rainforest	Sherbrook, shale	80	86	115	135
Eucalyptus saligna wet sclerophyll	Sherbrook, shale	43	63	65	102
Eucalyptus saligna wet sclerophyll	National Park, shale and sandstone	37	39	132	131
Eucalyptus pilularis wet sclerophyll	Turramurra, shale	33	37	91	113
Eucalyptus gummifera dry sclerophyll	National Park, sandstone	28	29	45	N.A.
Eucalyptus gummifera dry sclerophyll	Hornsby, sandstone	23	31	82	N.A.
Eucalyptus gummifera dry sclerophyll	Terry Hills, laterite	25	24	37	N.A.
Eucalyptus gummifera dry sclerophyll	National Park, laterite	13	19	93	130
Banksia–Acacia scrub	St. Ives, sandstone	16	22	19	N.A.

N.A.: not available.

Adapted from Beadle (1954).

growth response to the flax grown on rainforest soils, nitrogen and phosphorus had to be added to some soils collected from surrounding non-rainforest vegetation. However, flax grown on some other soils showed no response to these nutrients (Table 4.3). Beadle (1962*a*) acknowledged that pot trials might be misleading if edaphic factors other than soil fertility are primarily responsible for patterns that are observed in the field, and suggested that field experiments can potentially overcome these problems. The detailed field experiments about the

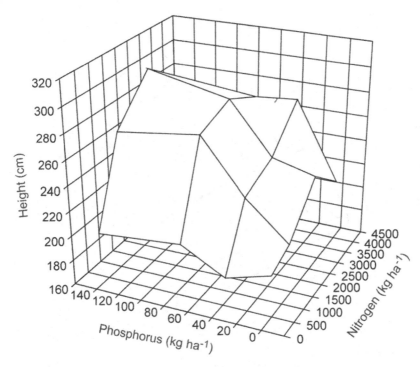

Figure 4.7
Mean height (cm) of 3.4-year-old Araucaria cunninghamii *grown under 16 different combinations of phosphorus and nitrogen fertilisation in plantations established on a site that had formerly supported dry* Eucalyptus *forest. Adapted from Richards and Bevege (1969).*

factors that control the establishment of the rainforest conifer *Araucaria cunninghamii* conducted by Richards (1967, 1968) and Richards and Bevege (1969) are therefore of great relevance to Beadle's hypothesis. They found that the addition of nitrogen, but not phosphorus, were critical in controlling the growth of *Araucaria cunninghamii* on soils that originally supported dry *Eucalyptus* forests (Figure 4.7). Indeed, Figure 4.7 shows that at the lowest levels of nitrogen supplementation there was little increment in *Araucaria cunninghamii* height in response to even maximum phosphorous supplementation. The multiple regression analysis of Richards and Bevege (1969) showed that foliar nitrogen, not phosphorus, accounted for most of the variation in the height growth of *Araucaria cunninghamii* (Figure 4.8). Without nitrogen supplementation, seedlings made 'virtually no growth at all, although they survive in a chlorotic condition for many years' (Richards 1968). Richards (1968) noted that when nitrogen deficiency was corrected in *Eucalyptus* soils, growth rates of *Araucaria*

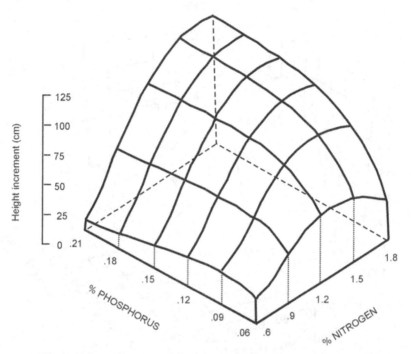

Figure 4.8

Relationship between foliar nitrogen and phosphorus and the height of growth of the rainforest conifer Araucaria cunninghamii *grown in plantations under 16 different levels of nitrogen and phosphorus fertiliser treatments on* Eucalyptus *forest soils (see Figure 4.7). Multiple regression analysis revealed that 62% of the height increment was accounted for by N, 14.7% by P and 4.2% by the interaction of N and P. Adapted from Richards and Bevege (1969).*

cunninghamii in plantations were comparable to those of plantations on former rainforest sites.

Richards realised that field experiments with *Araucaria cunninghamii* falsified Beadle's hypothesis, and attempted to reconcile this conflict by suggesting that his results may be ultimately compatible with Beadle's soil phosphorus hypothesis if soil nitrogen concentrations are directly governed by the availability of soil phosphorus. However, this suggestion is not supported by his demonstration of a limited growth response of *Araucaria cunninghamii* to the addition of phosphorus alone (Figure 4.7).

Conclusion

The evidence presented in this chapter does not support the idea that rainforests are necessarily restricted to soils with high phosphorus levels. Indeed, published data show that rainforests occur across an extraordinarily wide range of soil phosphorus. This pattern mirrors the wide range of foliar phosphorus concentrations reported for rainforest species in the previous chapter (Figure 3.11). Some studies show abrupt changes in phosphorus concentration across rainforest boundaries on uniform parent materials. This pattern raises the issue of cause and effect. Are high concentrations of soil nutrients, such as phosphorus, a consequence of distinct nutrient cycling patterns controlled by other ecological factors such as moisture supply, fire, drainage, topography or soil texture (Figures 4.3 and 4.4)? Aspects of these questions will be considered in Chapter 9 and the following two chapters.

5

The edaphic theory II. Soil types, drainage, and fertility

Andrews (1916) eloquently expressed the idea that soil fertility was critical in delimiting rainforest when he wrote that around Sydney:

there are sheltered areas or pockets of volcanic soil or shales, on which dense luxuriant plant growths abound forming canopies of dark and glossy green, which exclude a great proportion of the sun's rays. Surrounding these patches are the hungry sandstones forming so much of the large Sydney and Blue Mountains district, whose vegetation is in striking contrast with that of the rich soils. Here is to be seen no luxuriant foliage, no twining nor towering canopy to the jungle, but instead merely an array of *Eucalyptus*, phyllodineous acacias, banksias of sombre hue and casuarinas.

Subsequently, a number of ecologists have also advocated the idea that soil fertility is critical in determining the distribution of rainforest vegetation (McLuckie and Petrie 1927; Davis 1936; Fraser and Vickery 1938, Pidgeon 1940; Herbert 1960; Florence 1964, 1965; Webb 1968). The object of this chapter is to show that Australian rainforests are not consistently limited by soil texture, drainage, or soil fertility.

Geology and soil types

While it is true that the largest tracts of rainforest in Australia are, and were before being cleared, associated with fertile basaltic soils, the importance of soil parent material in controlling rainforest has been over-emphasised. Throughout Australia, rainforests are known to occur on a huge range of parent rock types including quartzite, sandstone, limestone, granite, basalt, rhyolite, granodiorite and laterite (Jackson 1968; Webb 1968; Ash 1988; Russell-Smith 1991; Fensham

1995). At a more local scale, heterogeneity of parent rocks was clearly demonstrated by Ash (1988) for the Atherton Tablelands, who showed that rainforests occurred on a range of rock types including acid and basic volcanics, granite and metamorphics notwithstanding the presence of large areas of Tertiary and Quaternary basalt flows. Further, Ash found that only 38% of the rainforest boundaries corresponded with a major geological disjunction.

Mirroring the huge geographic range and great variety of geologies is the wide range of different soil types that support rainforest. For example, Howard and Ashton (1973) noted that, in Victoria, *Nothofagus cunninghamii* forests occur on kraznozems, alpine humus soil, grey loams and colluvial and alluvial soils. The relative importance of physical and chemical properties associated with these different soil types will now be considered.

Soil physical properties

There is no evidence that rainforest is necessarily restricted to soils with specific textural properties. Indeed, a feature of the published data, especially when considered in aggregate, is the extremely wide range of physical properties of rainforest soils. Three examples are provided to substantiate this view.

i. A survey of sites on the east coast of Australia showed that there was no clear relationship between non-rainforest and rainforest formations and 10 surface and sub-surface soil physical characteristics (Tracey 1969) (Table 5.1). A two-dimensional ordination of sites based on these physical characteristics failed to discriminate between rainforest and non-rainforest vegetation (Figure 5.1). The vertical axis of the ordination was interpreted by Tracey as defining a gradient from poorly structured sandy soils to well structured clay soils, while the horizontal axis defined a gradient in the aeration and moisture content of soils (Table 5.1).

ii. Le Brocque and Buckney (1994) found that surface soil field capacity, saturation moisture content, sand, and silt-clay did not systematically differentiate *Ceratopetalum apetalum* rainforest on shale parent material from 10 other non-rainforest sites on both shale and sandstone substrates near Sydney (Table 5.2).

iii. Bowman (1992*a*) showed that the sand, silt and clay content of surface soils across 97 dry monsoon rainforest–savanna boundaries and 47 wet monsoon rainforest–savanna boundaries were not significantly different (Table 5.3). Surface soils of savanna communities had a significantly greater gravel content than adjacent wet monsoon rainforests, although the coefficients of variation for gravel content were high.

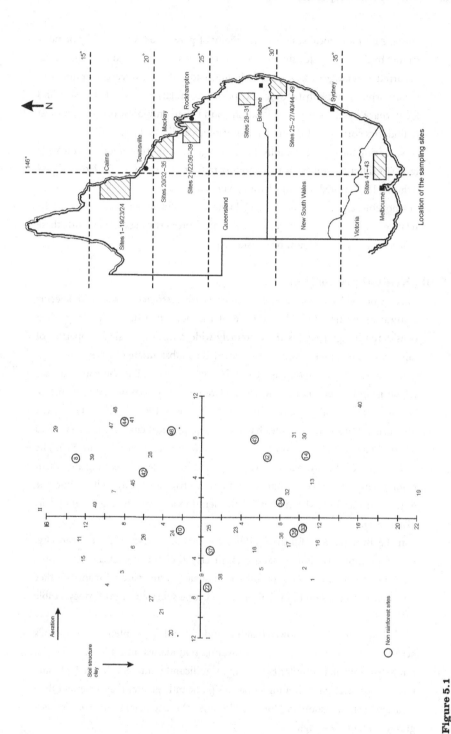

Figure 5.1

Ordination of rainforest (uncircled) and non-rainforest (circles) sites based on ten physical properties of their soils. The map shows the six regions on the east coast of Australia containing the 49 sites. Spearman rank correlation coefficients between the axis scores and the ten soil variables are presented in Table 5.1. Adapted from Tracey (1969).

TABLE 5.1. Spearman rank-correlation coefficients between 10 soil
physical variables and the Gower ordination of rainforest and
non-rainforest soils collected from the east coast of Australia

Soil physical variable	Axis I	Axis II
Bulk density at 0.01 MPa suction (g cm^{-3})	− 0.326	− 0.885
Particle density (g cm^{-3})	− 0.430	− 0.048
Total pore space at 0.01 MPa (% volume)	0.286	0.922
Water content at 0.01 MPa (% oven-dry weight)	0.155	0.913
Macroporosity at 0.01 MPa (% volume)	0.620	0.309
Bulk density at 1.5 MPa (g cm^{-3})	− 0.414	− 0.792
Water content at 1.5 MPa (% oven-dry weight)	− 0.065	0.914
Pore space at 1.5 MPa suction (% volume)	0.402	0.824
Air-filled pores at 1.5 MPa suction (% volume)	0.895	− 0.133
Available water (% volume)	0.630	− 0.411

Adapted from Tracey (1969).

TABLE 5.2. Soil physical characteristics from a *Ceratopetalum apetalum*
rainforest and the lowest and highest values from ten non-rainforest
sites in Ku-ring-gai Chase National Park
Values are means, (standard errors) and letters denote means that are not
significantly different ($P = 0.05$, Tukey multiple comparisons tests). The
non-rainforest sites supported open scrub/closed heath, tall shrubland/open
heath, woodland, woodland/open forest, open forest and tall woodland/tall open
forest.

		Non-rainforest types	
Soil physical characteristic	Rainforest	Minimum	Maximum
Gravel (% oven-dry weight)	4.2 (4.0) ab	1.0 (1.0) a	12.3 (10.0) b
Sand (% oven-dry weight)	77.6 (9.7) a	41.4 (21.9) b	84.5 (5.0) a
Silt/clay (% oven-dry weight)	18.5 (7.3) a	14.9 (4.5) a	45.3 (17.7) b
Saturation moisture content (% oven-dry weight)	36.7 (8.1) ab	31.7 (4.7) b	44.3 (2.4) a
Field capacity (% oven-dry weight)	32.9 (10.0) a	27.6 (4.7) a	41.6 (1.9) b

Adapted from Le Brocque and Buckney (1994).

TABLE 5.3. Soil texture variables from monsoon rainforests and
adjacent savannas collected on 144 rainforest boundary transects
throughout northwestern Australia

Values are means (coefficient of variation)

Soil variable	Wet monsoon rainforest boundary		Dry monsoon rainforest boundary	
	rainforest	savanna	rainforest	savanna
Coarse sand (%)	45.2 (52)	45.4 (47)	41.1 (50)	44.1 (40)
Fine sand (%)	38.0 (49)	37.0 (39)	36.8 (46)	34.8 (34)
Silt (%)	7.7 (77)	9.1 (84)	9.8 (80)	9.1 (64)
Clay (%)	9.1 (99)	8.6 (65)	12.3 (51)	12.2 (66)
Gravel (%)	3.3 (262)	13.4 (155)	19.0 (120)	15.6 (130)

Adapted from Bowman (1992a).

Soil drainage

Soil drainage is certainly responsible for controlling the spatial distribution of
particular Australian rainforest tree species (Turner and Kelly 1981; Bowman
and McDonough 1991; Melick and Ashton 1991). There is less evidence to
suggest that rainforest vegetation is limited by perennial or periodic waterlogging
of soils. Webb (1968) suggested that on sites with impeded drainage in the
Australian seasonal tropics and subtropics, *Melaleuca* or *Eucalyptus* forest or
sedge savannas replace rainforest. In the monsoon tropics, Bowman and
McDonough (1991) demonstrated the occurrence of such a transition. They
found that wet monsoon forest was replaced by *Melaleuca* forest where wet
season floodwater exceeded 1 m in depth, and that the *Melaleuca* forest was in
turn replaced by treeless sedgeland where floodwaters exceeded 2 m depth
(Figure 5.2). However, it is unclear if the patterns were solely determined by
floodwater depth or if the boundaries were also controlled by recurrent fires that
originated on the treeless floodplains (Bowman and Rainey 1995). Indeed, the
occurrence of small patches of rainforest amongst seasonally flooded sedgelands
may be a stage in the colonisation of the floodplains by rainforest (Bowman
1992a). Russell-Smith (1991) and Bowman (1992a) found that wet monsoon
rainforest occurred on perennially moist soils associated with springs, seeps,
creek banks and canyon floors. These sites are often flooded each wet season and
remain waterlogged throughout the dry season.

Figure 5.2
Monsoon rainforest on the edge of a large seasonally flooded sedgeland in the Australian monsoon tropics. (Photograph: David Bowman.)

Many rainforests in southern Australia are also restricted to the margins of rivers and creeks (Ashton and Frankenberg 1976, Gibson *et al.* 1991; Melick and Ashton 1991) and these gallery forests are flood-prone. Indeed, Melick and Ashton (1991) described riverine rainforests in Victoria which were subject to extensive flooding as a result of the river rising up to 14 m above normal (Figure 1.17)! As will be discussed in the following chapters, the occurrence of gallery rainforests may reflect protection from wildfire rather than implying a direct ecological dependence upon specific hydrological conditions.

Soil fertility

A large number of studies have compared the fertility of rainforest soils with non-rainforest soils. In aggregate, these studies show that rainforest occurs across a wide range of soil fertility conditions, and although rainforests are often on soils of greater fertility than surrounding non-rainforest vegetation, this is not always the case. I provide four case studies to substantiate this view.

i. Webb (1969) carried out an analysis of nine chemical characteristics of the surface soil at 22 rainforest and five non-rainforest sites in three regions on the east coast of Australia (Cairns, Fraser Island and Port Macquarie; Figure 5.3). A numerical classification of the sites on their surface soil chemical characteristics failed to discriminate between rainforest and non-rainforest vegetation

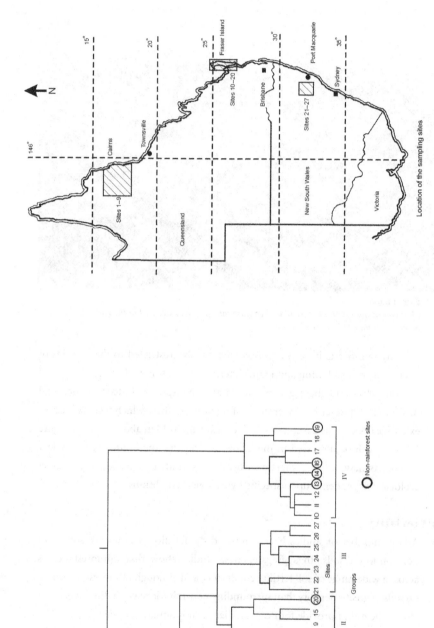

Figure 5.3

Classification of rainforest (uncircled) and non-rainforest (circled) sites based on nine chemical properties of their surface soils (see Table 5.4). The map shows the three regions on the east coast of Australia containing the 27 sites. Adapted from Webb (1969).

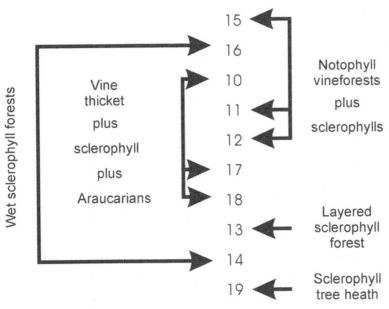

Figure 5.4
Uni-dimensional ordination of soil chemical data collected from a range of forest types on Fraser Island. The Spearman rank correlation coefficients between the axis scores and chemical variables are given in Table 5.5. Adapted from Webb (1969).

(Figure 5.3 and Table 5.4). Although most of the non-rainforest sites were placed in the classificatory Group IV that had the lowest mean concentration of most of the measured characteristics, this group also included five rainforest sites. Classificatory Group II, which had the highest levels of most measured characteristics, included one non-rainforest site, a sedge swamp from Fraser Island.

Webb (1969) conducted a more detailed analysis of the relationship between soil chemistry and forest distribution for the samples in Group IV. With the exception of one site from the mainland, all samples in Group IV occurred on Fraser Island (Figure 5.3). Fraser Island is a large coastal island in southeast Queensland with deep infertile siliceous soils derived from coastal sands that overlie Cretaceous sedimentary rock. Webb found that there was a strong gradient in the fertility of surface soils sampled from various forest types as defined by a uni-dimensional ordination of the soil fertility data (Figure 5.4, Table 5.5). In general, subtropical rainforest was restricted to the high-fertility sites and *Eucalyptus* forests to the low-fertility sites. One exception was a forest dominated by *Eucalyptus pilularis* that occurred on soils chemically similar to those that supported subtropical rainforests. Given that the parent material, deep sands, occurred uniformly across the gradient, it is reasonable to assume that the

TABLE 5.4. Surface soil chemical characteristics (means) for four groups of sites supporting rainforest and non-rainforest vegetation on the east coast of Australia (Figure 5.3). The 27 sites were allocated to groups following a numerical classification of the sites according to soil chemical data.

	Group I	Group II	Group III	Group IV
Number of sites	5	6	7	9
Number and types of non-rainforest vegetation	nil	1 sedge swamp	nil	2 wet sclerophyll *Eucalyptus* forests
				2 dry sclerophyll *Eucalyptus* forests
pH	4.7	5.2	N.A.	5.7
N (%)	0.213	0.439	0.364	0.051
Total PO_4 (ppm)	633	2501	1556	90
Available PO_4 (ppm)	N.A.	N.A.	218	18
Exchangeable K mEq 100 g^{-1} dry soil	0.27	0.50	0.63	0.14
Exchangeable Na mEq 100 g^{-1} dry soil	0.550	0.537	0.198	0.058
Exchangeable Ca mEq 100 g^{-1} dry soil	0.52	4.15	2.16	2.11
Exchangeable Mg mEq 100 g^{-1} dry soil	0.55	2.56	1.50	0.89
Exchangeable H mEq 100 g^{-1} dry soil	6.62	4.62	21.16	6.17

N.A.: not available.

Adapted from Webb (1969).

TABLE 5.5. Spearman rank-correlation coefficients between nine soil chemical variables and the first axis of the Gower ordination of rainforest and non-rainforest soils collected on Fraser Island

Chemical variable	Surface soil (0–10 cm)	Sub-surface soil (25–30 cm)
pH	− 0.018	− 0.176
N (%)	0.915	0.879
Total PO$_4$ (ppm)	0.721	− 0.152
Available PO$_4$ (ppm)	0.684	0.576
Exchangeable K mEq 100 g^{-1} dry soil	0.891	0.322
Exchangeable Na mEq 100 g^{-1} dry soil	0.709	0.527
Exchangeable Ca mEq 100 g^{-1} dry soil	0.939	0.309
Exchangeable Mg mEq 100 g^{-1} dry soil	0.612	− 0.030
Exchangeable H mEq 100 g^{-1} dry soil	0.067	− 0.756

Adapted from Webb (1969).

cause of the soil fertility gradient was differences in the accession of nutrients. Nutrients would be transported by wind and water, which are in turn controlled by the interaction of vegetation type, topography, site exposure and site fire history (Webb 1969). Clearly, the fertile rainforest soils have developed independently of the siliceous parent material.

ii. In the Sydney region, Le Brocque and Buckney (1994) found that soils of *Ceratopetalum apetalum* rainforests on shale substrates were not significantly different from those of non-rainforest communities on shales and sandstones for any of the 13 chemical variables that they measured (Table 5.6).

iii. Turner and Kelly (1981) conducted a detailed survey of the soils of a small catchment on the north coast of New South Wales. The catchment contained several types of subtropical rainforest and sclerophyll forest. They found that soil nutrient concentrations varied significantly between forest types (Table 5.7). For example, *Eucalyptus pilularis* forest occurred on dry slopes with infertile soils, while *Eucalyptus grandis* forests with a dense rainforest understorey occurred on moist sites with soils of comparable nutrient levels to adjacent subtropical rainforest dominated by *Argyrodendron trifoliatum*, *Sloanea woollsii* and *Toona australis* (Table 5.7). The soils beneath *Ceratopetalum apetalum* rainforest, *Archontophoenix cunninghamiana* rainforest and *Lophostemon confertus* sclerophyll forest were all infertile. They suggested that the fertility differences they observed

TABLE 5.6. Soil chemical characteristics from a *Ceratopetalum apetalum* rainforest and the lowest and highest values from ten non-rainforest sites in Ku-ring-gai Chase National Park

Values are means, (standard errors) and notation of means that are not significantly different ($P = 0.05$, Tukey multiple comparisons tests). The non-rainforest sites supported open scrub/closed heath, tall shrubland/open heath, woodland, woodland/open forest, open forest and tall woodland/tall open forest.

		Non-rainforest	
Soil chemical characteristic	Rainforest	Minimum	Maximum
pH	5.6 (1.1) a	5.1 (0.2) b	6.5 (0.2) a
Electrical conductivity (μs cm^{-1})	11.2 (1.7) a	10.8 (1.9) a	21.1 (12.5) b
Loss on ignition (% oven-dry weight)	8.5 (3.0) a	2.3 (0.3) b	7.7 (1.4) a
Na mEq 100 g^{-1} dry soil	0.099 (0.036)	0.045 (0.014)	0.168 (0.050)
K mEq 100 g^{-1} dry soil	0.218 (0.118) a	0.047 (0.016) b	0.189 (0.093) a
Mg mEq 100 g^{-1} dry soil	0.412 (0.268) ab	0.069 (0.033) a	0.882 (0.630) b
Ca mEq 100 g^{-1} dry soil	0.344 (0.303) a	0.022 (0.002) a	1.829 (1.452) b
CEC mEq 100 g^{-1} dry soil	13.73 (3.78) a	4.84 (0.75) b	13.36 (3.02) a
P (% dry soil)	0.011 (0.004) a	0.004 (0.001) b	0.014 (0.030) a
N (% dry soil)	0.173 (0.059) a	0.036 (0.012) b	0.187 (0.045) a

Adapted from Le Brocque and Buckney (1994).

within the catchment were related to variation in soil parent material, with the exception of the rainforest dominated by *Ceratopetalum apetalum*, which is known to accumulate aluminium in its foliage and leaf litter (Webb 1954). This accumulation leads to high concentrations of aluminium in the surface soil, with consequent base depletion (Webb *et al.* 1969).

iv. Bowman (1992*a*) conducted a study of monsoon rainforest–savanna boundary soils throughout northwestern Australian. Wet monsoon rainforests had significantly higher electrical conductivity, total sulphur, total nitrogen, available sodium, and available potassium than surrounding savanna. The dry monsoon rainforest soils had significantly higher levels of most chemical variables compared to soils in the surrounding savanna. The exceptions were potassium, which was at higher concentrations in the savanna soils, and sodium, for which there was no difference (Table 5.8). A conspicuous feature of these data is the massive variation in the nutrient concentrations within and between the

TABLE 5.7. Chemical characteristics of surface soils (0–7.5 cm) collected from three types of subtropical rainforest and three types of non-rainforest that occur within a small catchment on the northeast coast of New South Wales

Forest type	No. of plots	pH	N (%)	OM (%)	Exchangeable cation (mEq %)					Total exchangeable bases	Total P (ppm)
					Al	Ca	Mg	K	Na		
Argyrodendron trifoliatum, Sloanea woollsii and Toona australis subtropical rainforest	9	4.81	0.81	18.1	1.18	9.30	3.43	1.08	0.37	14.18	1340
Archontophoenix cunninghamiana rainforest	4	4.40	0.45	9.2	2.66	2.14	2.66	0.71	0.18	5.70	370
Ceratopetalum apetalum rainforest	5	3.80	0.69	28.1	7.95	1.14	2.15	0.90	0.64	4.85	340
Eucalyptus pilularis forest	6	4.95	0.37	15.9	3.05	4.50	3.45	0.74	0.34	9.14	195
Eucalyptus grandis forest	7	4.72	0.64	17.7	1.40	5.83	3.38	1.18	0.38	10.77	1095
Lophostemon confertus sclerophyll forest	7	4.45	0.57	23.9	2.88	3.56	2.70	0.78	0.52	6.85	435
Least significant difference $P < 0.05$		0.59	0.29	10.2	N.A.	N.A.	N.S.	N.S.	N.S.	N.A.	N.A.
Least significant difference $P < 0.01$		N.A.	N.A.	N.A.	1.84	1.90	N.A.	N.A.	N.A.	2.80	360

N.A.: not available; N.S.: not significant.
OM: organic matter.
Adapted from Turner and Kelly (1981).

TABLE 5.8. Chemical attributes of surface soils on either side of 144 monsoon forest/savanna boundaries in northwestern Australia.

Values are means, (coefficients of variation) and notation of means that are not significantly different ($P > 0.05$) as determined using Wilcoxon tests (wet and dry monsoon forest boundaries) and t-tests (*Allosyncarpia ternata* rainforest boundaries)

Chemical variable	Wet monsoon forest boundary		Dry monsoon forest boundary		*Allosyncarpia ternata* rainforest boundary	
	rainforest	savanna	rainforest	savanna	rainforest	savanna
pH	5.3 (19) a	5.5 (14) a	6.1 (16) b	6.0 (16) b	4.5 (4) c	4.5 (3) c
Electrical conductivity (μs cm^{-1})	168.4 (74)	76.4 (66)	269.5 (118)	131.6 (80)	27.4 (13) c	24.3 (12) c
Total P (%)	0.027 (118) a	0.018 (76) a	0.089 (279)	0.036 (111)	58 (24) c	49 (22) c
Total K (%)	0.093 (212) a	0.097 (228) a	0.365 (194)	0.442 (233)	260 (12) c	257 (13) c
Total S (%)	0.054 (139)	0.021 (53)	0.065 (122)	0.026 (71)	40 (28) c	36 (51) c
Total C (%)	2.754 (66) a	2.239 (77) a	2.861 (64)	2.140 (59)	0.288 (43) c	0.300 (34) c
Total N (%)	0.230 (110)	0.121 (83)	0.284 (101)	0.139 (71)	0.024 (59) c	0.018 (59) c
Available Na (mEq 100 g^{-1})	0.190 (80)	0.010 (84)	0.359 (572) b	0.116 (74) b	0.0288 (39) c	0.0300 (26) c
Available K (mEq 100 g^{-1})	0.171 (111)	0.099 (125)	0.442 (120)	0.314 (130)	0.0213 (53)	0.0114 (23)
Available Ca (mEq 100 g^{-1})	7.984 (154) a	4.998 (172) a	16.073 (133)	7.568 (118)	0.285 (64) c	0.233 (35) c
Available Mg (mEq 100 g^{-1})	5.455 (145) a	1.862 (141) a	3.179 (116)	2.006 (102)	0.149 (32) c	0.134 (28) c
Cation exchange capacity (mEq 100 g^{-1})	13.76 (145) a	7.060 (148) a	20.05 (121)	10.01 (106)	0.484 (48) c	0.406 (27) c
No. of samples	59	33	125	57	8	7

Adapted from (Bowman 1991*a*, 1992*a*).

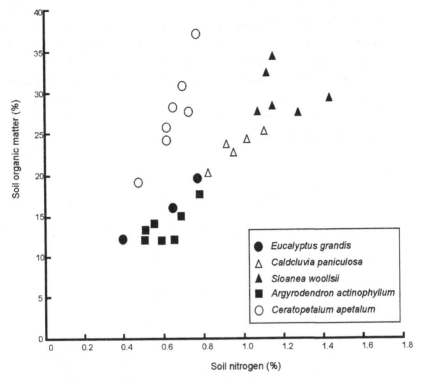

Figure 5.5
Relationship of four subtropical rainforest tree species and Eucalyptus grandis *with total concentrations of nitrogen and soil organic matter in the surface soil beneath them. Adapted from Lambert* et al. *(1983) and reproduced with permission of CSIRO Australia.*

different vegetation types: both savanna and rainforest occurred on extremely infertile soils. This is well illustrated by the detailed study of a dry monsoon rainforest dominated by the myrtaceous tree *Allosyncarpia ternata* that is endemic to the Arnhem Land Plateau (Bowman 1991a). *Allosyncarpia* rainforest occurred on deep, excessively drained siliceous soils derived from Proterozoic sandstone. Comparisons of the chemistry of surface soils of the *Allosyncarpia* rainforest and the adjacent savanna revealed that only exchangeable potassium was significant different (Bowman 1991a) (Table 5.8).

Conclusion

The evidence presented in this chapter does not support the generalisation that rainforest and non-rainforest vegetation occupy sites that differ consistently in the physical and chemical characteristics of their soils. This conclusion, however, does not belittle the importance of soil fertility as a correlate with the distributional patterns of both rainforest and non-rainforest tree species and different types of rainforest and non-rainforest vegetation (Lambert *et al.* 1983; Lambert and Turner 1983) (Figure 5.5). But such correlations do not imply causal relationships. This chapter has ignored the evidence that complex interactions between vegetation, fire, topography and edaphic factors may be important in determining the formation of rainforest soils (Webb 1968, 1969; Ash 1988; Melick 1990*a*; Le Brocque and Buckney 1994). The interaction of these variables will be considered in Chapter 9.

6

The climate theory I. Water stress

Schimper's (1903) original use of the term 'rainforest' succinctly encapsulated his belief that luxuriant, floristically and structurally complex tropical vegetation could only develop in a drought-free climate. Although some Australian ecologists have adhered to the idea that moisture supply is of critical importance in controlling the distribution of Australian rainforest (Specht 1981a), there are several features of Australian rainforests that are inconsistent with this theory. Firstly, there is an imperfect correlation between high rainfall areas and the distribution of rainforest, a fact that puzzled pioneering ecologists such as McLuckie and Petrie (1927). Secondly, in the latter half of this century many Australian ecologists broadened the concept of rainforest to include vegetation that occurs in low rainfall and seasonally dry climates (Webb 1959; Baur 1968; Webb and Tracey 1981; Adam 1992) (Figure 6.1). This broadening of definition is reflected in the use of the oxymoron 'dry rainforest' (Baur 1957; Adam 1992; Fensham 1995). Indeed, many authors have opted to spell rainforest as a single word (rather than the more conventional 'rain forest') in order to play down the implication that high rainfall is critical in controlling the distribution of rainforest in Australia (Baur 1968; Webb and Tracey 1981; Adam 1992). Thirdly, there is evidence that some Australian rainforest trees, including some that occur in high rainfall environments, are extremely drought tolerant, being able to endure seasonal moisture deficits with 'leaf water potentials more commonly associated with arid environments' (Yates et al. 1988). The purpose of this chapter is to demonstrate that the available evidence does not support the theory that the geographic distribution of rainforest is solely explained by seasonal water stress.

Figure 6.1
Interior of a 'dry rainforest' on the Robinson River in New South Wales. (© Murray Fagg, Australian National Botanical Gardens.)

Rainforests and moisture gradients

Webb (1968) and Webb and Tracey (1981) suggested that Australian tropical rainforests form a sequence along a precipitation gradient and are ultimately replaced by *Eucalyptus* savanna at around 600 mm annual rainfall. This threshold is broadly consistent with the observation of Fensham (1995) that in northern Queensland the western limit of rainforest is about 280 km from the coast where rainfall is about 500 mm per annum. Russell-Smith (1991) also found that the arid limit of rainforests in the Northern Territory was about 600 mm per annum (Figure 6.2). In Victoria, Melick (1990a) described rainforests at their geographic limit in a region that received about 700 mm per annum. Along decreasing moisture gradients there is a marked reduction in the productivity of rainforest trees. For example, Unwin and Kriedemann (1990) found that the growth of both *Melia azederach* (a deciduous species) and *Acacia aulacocarpa* (an evergreen species) declined along a moisture gradient from 2800 to 760 mm per annum in north Queensland (Table 6.1).

It is difficult to determine the precise arid limit of rainforest for three reasons. First, rainforests that occur on springs and sheltered canyons are effectively protected from the regional climate (Russell-Smith 1991; Bowman *et al.* 1991a). An extreme example of this is the occurrence of *Livistona* palms along perennial

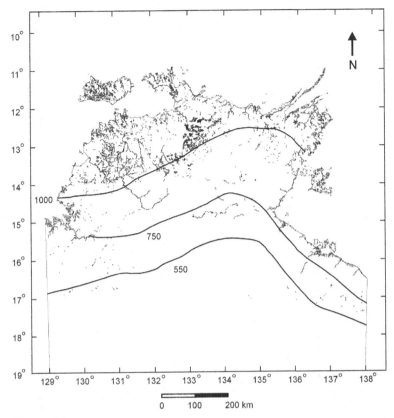

Figure 6.2

Distribution of monsoon rainforest vegetation and mean annual rainfall (mm) in the Northern Territory. Adapted from unpublished aerial photographic interpretation by Russell-Smith and Lucas.

seepages in Palm Valley in central Australia, where the mean annual rainfall is about 300 mm (Latz 1975). Second, some authors have specifically included or excluded some arid vegetation types in their definition of rainforest. For example, Russell-Smith (1991) opted to include *Acacia shirleyi* shrublands as a type of Northern Territory monsoon rainforest, but excluded *Acacia aneura* shrublands that are widespread in central Australia, whereas Fensham (1995) specifically excluded *Acacia* shrublands from his definition of 'dry- rainforest' in north Queensland. If *Acacia* shrublands are included in the definition of rainforest then there is no clear relationship between rainfall and the distribution of rainforest because vast tracts of *Acacia* shrublands occur in the arid regions of central Australia (Figure 6.3). Third, there is no consensus as to what constitutes a 'rainforest' species. It is well known that some species that occur in rainforests have wide ecological ranges as evidenced by their occurrence in non-rainforest,

TABLE 6.1. Site characteristics and variation in leaf xylem water potential (ψ_l), seasonal leaf conductance (g_s), growth rate and deciduousness of *Acacia aulacocarpa* and *Melia azederach* at three sites along a moisture gradient

Site	Coastal		Subcoastal		Inland	
Distance to coast (km)	5		52		165	
Altitude (m)	80		720		780	
Mean annual rainfall (mm)	2800		1410		760	
Species	*Acacia aulacocarpa*	*Melia azederach*	*Acacia aulacocarpa*	*Melia azederach*	*Acacia aulacocarpa*	*Melia azederach*
Dry season minimum ψ_l (MPa)	−2.1	−1.7	−1.7	−1.8	−6.4	−2.0
Wet season minimum ψ_l (MPa)	−2.0	−2.0	−1.9	−2.0	−2.3	−2.3
Dry season mean daily max g_s (cm s^{-1})	0.3	0.3	0.2	0.2	0.05	0.1
Wet season mean daily max g_s (cm s^{-1})	0.6	0.68	0.4	0.52	0.2	0.47
Current annual increment in basal area (cm^2 year^{-1})	54.1	41.8	56.9	23.4	3.1	11.1
Months deciduous	0	2.5	0	3	0	4

Adapted from Unwin and Kriedemann (1990).

Figure 6.3
*Brigalow (*Acacia harpophylla*) shrubland with an emergent bottle-tree (*Brachychiton rupestris*) near Tambo in central Queensland. According to some definitions this vegetation could be considered the arid extreme of Australian rainforest. (© Murray Fagg, Australian National Botanical Gardens.)*

and even arid land, vegetation (Herbert 1960; Beadle 1966; Stocker and Mott 1981; Sluiter 1992; Bowman and Latz 1993). Species like *Acacia aulacocarpa* and *Acacia melanoxylon* are sometimes components of rainforests in humid environments in eastern Australia yet they occur across large rainfall gradients (Farrell and Ashton 1978; Unwin and Kriedemann 1990). Figure 6.4 shows that although the number of species classified by Liddle *et al.* (1994) as Northern Territory monsoon rainforest species declines sharply at the 600 mm isohyet, a small proportion persist in desert regions with less than 300 mm of annual rainfall. The intermixing of arid and rainforest tree species also points to the tolerance of many rainforest trees to extreme water stress. Ecophysiological studies clearly demonstrate that *Acacia harpophylla* (Connor and Tunstall 1968; Tunstall and Connor 1975, 1981), *Acacia aulacocarpa* (Unwin and Kriedemann 1990), *Callitris glaucophylla* (Clayton-Greene 1983; Attiwill and Clayton-Greene 1984) and *Callitris intratropica* (Bowman 1991*a*) are extremely tolerant of water stress. It is reasonable to assume that rainforest species that grow with these species in arid environments are also extremely drought tolerant.

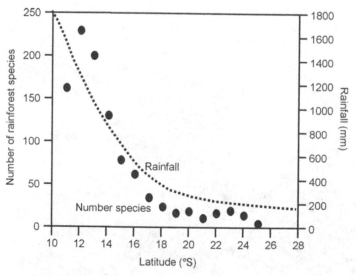

Figure 6.4

Relationship between latitude, mean annual rainfall and the number of rainforest species (as defined by Liddle et al. 1994) in grid cells (3.5° of longitude by 1° of latitude) that formed a continuous transect through the centre of the Northern Territory. For more details of the transect see Bowman and Connors (1996).

Poor correlation between rainfall and rainforest

Even within high rainfall areas, the distribution of rainforest is often patchy. McLuckie and Petrie (1927) found the abrupt boundaries between rainforest pockets and *Eucalyptus* forests at Mt Wilson in New South Wales 'remarkable' given the uniformity of the climate in that region.

The poor correlation between the distribution of *Nothofagus cunninghamii* and rainfall is provided to illustrate this point. Busby (1986) undertook an analysis of the relationship between the geographic distribution of *Nothofagus cunninghamii* and long-term climatic averages, and concluded that this species was essentially in equilibrium with prevailing climates. He suggested that mean summer (January–March) precipitation of 170–190 mm was most strongly related to the spatial distribution of the species, with the exception of an area in northeastern Victoria which could reflect the impact of past unfavourable climates (Figure 6.5). The generalisation that in Victoria this species is in equilibrium with current climates breaks down at finer spatial scales, as *Nothofagus cunninghamii* is restricted locally, typically to gullies (Howard and Ashton 1973; Figure 6.5).

Jackson (1968) also noted the importance of high summer rainfall in controll-

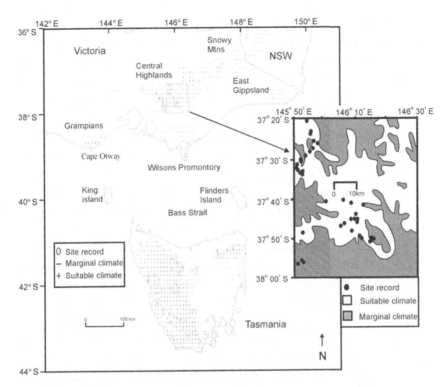

Figure 6.5

Site records of Nothofagus cunninghamii *and predicted distribution of 'potentially suitable'
and 'potentially marginal' climates for the species based on six air temperature and six
precipitation variables. A 'potentially suitable' climate is one where the values of six air
temperature and six precipitation variables used in spatial modelling were all within the 90
percentile range of sites that supported* N. cunninghamii. *'Potentially marginal' climates are
those where all six air temperature and all six precipitation variables were within the recorded
range for* N. cunninghamii *but one or more was outside the 90 percentile range. Unsuitable
climates are those for which any one of the 12 climate variables was outside the range known
to support* N. cunninghamii. *The insert shows site records of* N. cunninghamii *and the
distribution of 'potentially suitable' and 'potentially marginal' climates in part of central
Victoria. Adapted from Busby (1986).*

ing the distribution of *Nothofagus cunninghamii* in Tasmania. However, he
argued that the significance of summer rainfall was not in the physiological limits
of *Nothofagus*, as he believed the species to be 'surprisingly tolerant of water-
stress', but through its influence on the risk of fire. Busby (1986) acknowledged
that the biogeoclimatic model he used could not take into account factors such as
topography and its influence on local climate and fire risk. The role of fire in
controlling rainforest distribution will be discussed in Chapters 8, 9 and 10.
Jackson's (1968) assertion that *Nothofagus cunninghamii* is drought tolerant is
supported by field and greenhouse studies on seedlings conducted by Howard

(1973b). Seedlings grown in a glasshouse showed a remarkable tolerance to the cessation of watering, with some 2-year-old seedlings (average height 30–45 cm) surviving for 7 weeks before dying. During a severe drought, Howard compared the survival of 7-month-old *Nothofagus cunninghamii* seedlings planted in a moist *Eucalyptus nitens* forest, a dry *E. obliqua* forest and a *Nothofagus cunninghamii* forest. Seedling death only occurred in the *E. obliqua* forest where 85% of the initial 20 seedlings perished. Howard concluded that periodic severe droughts such as may occur once or twice a century could prevent *N. cunninghamii* from colonising dry *E. obliqua* forests but could not prevent its establishment in wet *Eucalyptus* forests. She suggested that in addition to moisture stress, periodic fires and herbivory may also limit the expansion of *Nothofagus cunninghamii* into *Eucalyptus* forests.

It must be admitted that the poor correlation of rainforest distribution with rainfall patterns may be related to soil fertility and topography. A number of authors have suggested that fertile soils permit rainforests to occur in drier climates than do infertile soils (Baur 1957; Webb and Tracey 1981; Ash 1988). Webb and Tracey (1981) have described this phenomenon as 'edaphic compensation'. Ash found clear evidence of this on the Atherton Tablelands in north Queensland, where rainforest occurs on basalt in areas with lower rainfall than it does on substrates with lower fertility (Figure 6.6). The reason for such 'edaphic compensation' is complicated and poorly understood, and it is possible that the role of soil fertility is related to ability of rainforests to recover from fire damage (Webb 1968). Nonetheless, rainforest is common in northern Australia on infertile substrates in low rainfall environments, showing that 'edaphic compensation' is not a universal phenomenon (Russell-Smith 1991; Bowman 1992a; Fensham 1995).

Topography can influence rainforest distribution, enhancing moisture supply locally by exposing sites to fogs or moisture laden winds and protecting sites from desiccating dry winds. Consider the asymmetric distribution of rainforests which occur on the Great Dividing Range in central New South Wales (Figure 6.7). Fisher (1985) interpreted the restriction of the rainforests to the southern side of the range as being related to the interception of moisture-bearing southeasterly winds. In the same region, Fraser and Vickery (1938) suggested that westerly winds might dry soil and hence limit rainforests to sites sheltered from these winds. Equally, the absence of rainforest on the northern side of the range could be related to exposure to fires driven by dry winds (Webb and Tracey 1981). Moisture held in low clouds can be an important source of water for rainforests (Howard 1973a; Ellis 1971a). Yates and Hutley (1995) provided

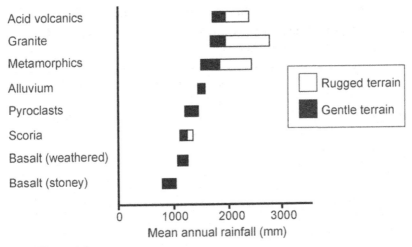

Figure 6.6
The geographic range of rainforest boundaries on the Atherton Tablelands defined by mean annual rainfall, substrate type and topography. Rainforest boundaries which occur on rugged topography are indicated by open bars. Adapted from Ash (1988).

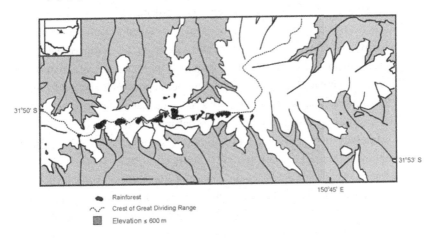

Figure 6.7
Distribution of rainforest on a section of the Liverpool Range, western central New South Wales highlands. Adapted from Fisher (1985).

experimental evidence that the subtropical rainforest tree *Sloanea woollsii* maintains positive water balance by absorbing fog and dew through its leaves during protracted rainless periods. Fog-drip can also locally increase soil moisture supply. For instance, Hutley *et al.* (1997) found that fog deposition provided 40% of the moisture reaching the floor of a subtropical rainforest in southeastern Queensland. There is some evidence that rainforest is more effective in trapping

TABLE 6.2. Number of days when the surface and sub-surface soil (10 and 48 cm depth) of three adjacent vegetation types fell below field capacity in northeastern Tasmania

	No. days less than field capacity	
Vegetation types	Summer 1964–65	Summer 1965–66
Nothofagus cunninghamii rainforest	0	28
Eucalyptus delegatensis forest	70	71
Grassland	102	71

Adapted from Ellis (1971*b*).

fog than non-rainforest vegetation (Ellis 1971*b*). For two consecutive summers, he showed that *Nothofagus cunninghamii* rainforest on a plateau in northeastern Tasmania had far fewer days when its surface soils fell below field capacity than adjacent *Eucalyptus delegatensis* forest and grassland (Table 6.2). Further, he suggested that spatial variation in fog frequencies might explain why rainforest understoreys develop beneath *Eucalyptus* forests on the northern edge of a plateau but not on its southern edge even though both sides receive a comparable amount of rainfall (Figure 6.8). The occurrence of grasslands and rainforests in Ellis's study area shows that there is not a clear-cut relationship between rainforest and landscapes with moisture supplies enhanced by low cloud.

Humidity is typically lower in non-rainforest vegetation than in adjacent rainforests (McLuckie and Petrie 1927; Myers *et al.* 1987). Myers *et al.* (1987) found no significant differences in a range of climatological variables between cleared and uncleared rainforest in the humid tropics with the exception of a greater number of hours when the relative humidity fell below 70% (Table 6.3). Leaves of the tropical rainforest canopy tree *Castanospermum australe* suffered greater water stress in the open environment than in the rainforest canopy (Figure 6.9). The greater water stress was accompanied by physiological differences at both full hydration and zero turgor, open-grown plants having lower osmotic potential in leaf tissue than canopy plants. Myers *et al.* (1987) suggested that physiological adjustment to microclimate may allow this species to colonise open environments. Fordyce *et al.* (1997*a*) demonstrated great differences in dry season atmospheric vapour pressure deficits between ravine and hill-top environments where *Allosyncarpia ternata* grew (Figure 6.10). The relationship between leaf-to-air vapour pressure difference (LAVPD) and transpiration rate

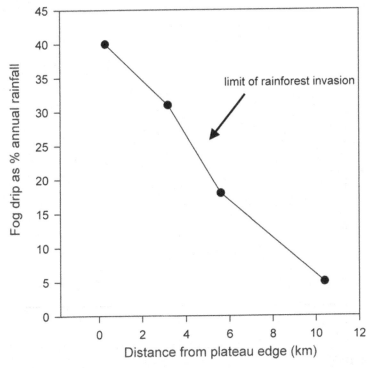

Figure 6.8

Average fog-drip, expressed as a percentage of annual rainfall recorded at four stations on an 850 m plateau in northeastern Tasmania. Sampling was conducted for two years. Along the gradient, the average annual rainfall for the two years varied from 1340 to 1510 mm, with about 17% falling in the summer months of December to March inclusive. Adapted from Ellis (1971b).

of *Allosyncarpia ternata* changed from a linear relationship at the beginning of the dry season to a uniformly low transpiration regardless of LAVPD at the end of the dry season. This result suggests that soil moisture supplies have overriding importance on transpiration rate relative to atmospheric humidity (Figure 6.11).

Soil moisture

Many workers have argued that soil moisture is of critical importance in determining the local distribution of rainforest (McLuckie and Petrie 1927; Pidgeon 1937; Fraser and Vickery 1939; Specht 1958; Florence 1964; Beard 1976; Specht *et al.* 1977; Hnatiuk and Kenneally 1981; Bowman and Dunlop 1986). The available data do not consistently support the view that rainforest occurs on soils with greater supplies of moisture than surrounding vegetation. Below I provide four case studies to substantiate this point.

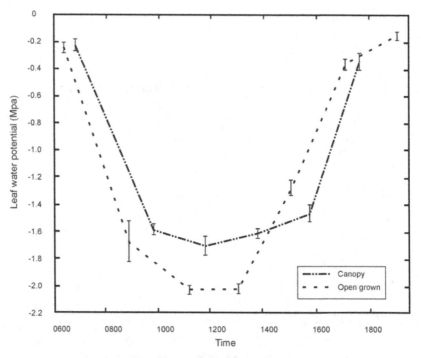

Figure 6.9

Mean (and standard error) of the leaf water potential of Castanospermum australe *trees growing in a large clearing and in the canopy of a rainforest on 28 July 1984. In each environment, three trees were sampled. The canopy trees were about 35 m tall and the open-grown trees were about 1.5 m tall. Micro-climatological data for the cleared and forest canopy environments are listed in Table 6.3. Adapted from Myers* et al. *(1987).*

i. The pioneering study of McLuckie and Petrie (1927) showed that a *Eucalyptus piperita* community on a sandstone plateau had very low levels of soil moisture compared with both *Eucalyptus goniocalyx* forests and *Ceratopetalum apetalum* rainforests on sandstone and basalt soils at Mt Wilson in the Blue Mountains (Figure 6.12). The latter communities were found to have overlapping ranges of soil moisture, especially where *Eucalyptus goniocalyx* forest had a dense fern understorey. These results do not support McLuckie and Petrie's conclusion that soil moisture is critical in determining the distribution of rainforest.

ii. On the central coast of New South Wales, Florence (1964) compared the variation in moisture supply over the course of 15 months of two *Eucalyptus* communities and a subtropical rainforest, all of which were growing on krasnozemic soils. For both of two winter dry seasons he sampled, the rainforest surface soil fell below wilting point, while the surface soil moisture levels in both

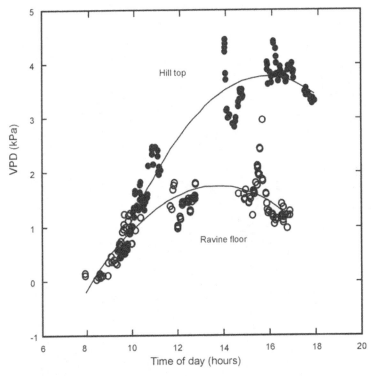

Figure 6.10

Comparison of the diurnal variation in vapour pressure deficit in an Allosyncarpia ternata *forest growing in a moist ravine and on a dry hill-top at the end of the dry season (November) in Kakadu National Park. Adapted from Fordyce et al. (1995).*

Eucalyptus communities only fell below wilting point in the second winter (Figure 6.13). The sub-surface soils of all three communities were consistently well below field capacity throughout the study period. The sub-surface soils in the *E. saligna* forest were below wilting point for all but a few months in the summer, while the *E. pilularis* and rainforest communities sub-surface soils only fell below their respective wilting points during the second winter sampling period. Florence assumed that rainforest trees are shallow rooted and therefore would be unaffected by the low levels of sub-soil moisture. He recognised that if this were not the case then the rainforest 'must be subject to moisture stress in most years and severe moisture stress in periodic drought years'. He rejected this conclusion, however, and formed the view that where 'other soil properties are not limiting' then 'soil moisture availability would be a critical factor' in the distribution of rainforest.

iii. Russell-Smith (1991) and Bowman *et al.* (1991*a*) argued that two

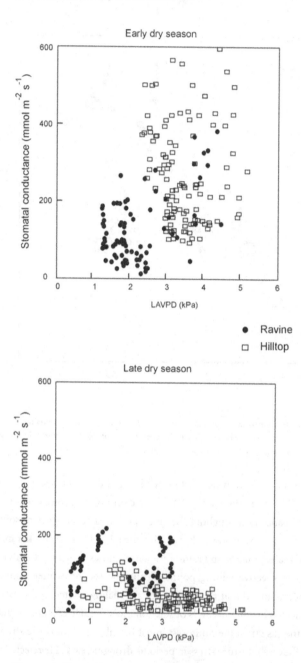

Figure 6.11
Variation between the early dry season and late dry season in the relationship between leaf-to-air vapour pressure difference (LAVPD) and stomatal conductance for Allosyncarpia ternata *leaves on trees growing in a moist ravine and on a dry hill-top. Adapted from Fordyce et al. (1997a).*

TABLE 6.3. Micro-climate of a cleared and rainforest canopy environment on the Atherton Tableland in northeast Queensland, July–August 1984

Values are means (standard error). At each site, leaf moisture stress of the tropical rainforest tree *Castanospermum australe* was also determined (Figure 6.9)

Micro-climatic parameter	Open-grown	Canopy
Daily total photosynthetically active radiation (mol m^{-2})	30.9 (1.7)	30.9 (1.7)
Daily maximum photon flux density (μmol m^{-2})	1776 (33)	1776 (33)
Daily maximum relative humidity (%)	98.3 (1.1)	99.7 (0.1)
Daily minimum relative humidity (%)	54.8 (6.1)	61.5 (3.3)
Time period where relative humidity < 70% (hours day^{-1})	4.7	3.0
Daily maximum temperature (°C)	20.9 (0.7)	22.2 (0.9)
Daily minimum temperature (°C)	12.6 (0.8)	12.5 (0.6)

Adapted from Myers *et al.* (1987).

Ceratopetalum–Doryphora forest (basalt).

Ceratopetalum–Doryphora forest (sandstone).

Eucalyptus goniocalyx forest: *Alsophila* and *Blechnum* understorey

Eucalyptus goniocalyx forest: *Pteridium* understorey

Eucalyptus piperita forest

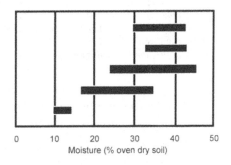

Figure 6.12
The range of surface soil moisture recorded in five forest types at Mount Wilson, Blue Mountains. Soils were collected at various times of year after periods of dry weather. Adapted from McLuckie and Petrie (1927).

monsoon rainforest types can be identified: those that occur on perennially moist substrates (e.g. springs, creek-lines and sheltered canyons), and rainforests which occur on well-drained and seasonally dry substrates (e.g. deep soils, cliff tops, boulder fields). Bowman (1992*a*) compared the dry season moisture content of the surface soil of wet and dry monsoon rainforests and surrounding savannas. He found that the moisture content of the surface soil of 44 wet

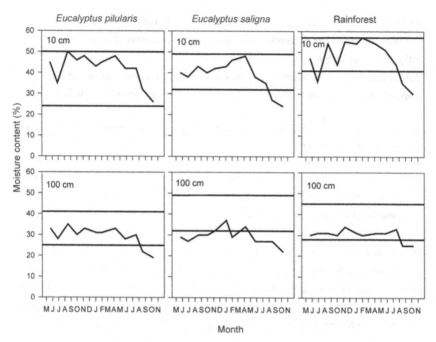

Figure 6.13

*Field capacity (upper horizontal line), wilting point (lower horizontal line) and moisture
content of surface (10 cm) and sub-surface (100 cm) soils for* Eucalyptus pilularis *forest,*
Eucalyptus saligna *forest, and subtropical rainforest dominated by* Argyrodendron
trifoliatum. *The soils were sampled over a 15-month period from the beginning of the winter
of 1959 until the end of spring in 1960. Adapted from Florence (1964).*

monsoon forests differed significantly from that of the surrounding wet savannas
(55.5 versus 9.3 g 100 g^{-1} of oven dry soil respectively). However, he found no
significant difference in the dry season surface soil moisture content of 66 dry
monsoon rainforests and that of their surrounding dry savannas (23.0 versus
11.9 g 100 g^{-1} of oven dry soil respectively). This conclusion is supported by the
more detailed study of Bowman and Panton (1993*b*). Figure 6.14 shows that
even though the dry monsoon rainforest surface soils had higher moisture
contents than adjacent savanna surface soils, the surface soils from both communities fell below wilting point at the end of the dry season.

iv. Howard (1973*b*) compared the surface and sub-surface moisture supply in
Eucalyptus nitens forest, *E. obliqua* forest and *Nothofagus cunninghamii* rainforest
during a severe drought in 1967–1968. She demonstrated that only the upper
2 cm of the *E. obliqua* soil surface fell below wilting point (Figure 6.15). Compared to the *Eucalyptus* communities, the soils at 2 cm depth in the *Nothofagus*
rainforest had greater moisture supplies in the early part of the summer (Decem-

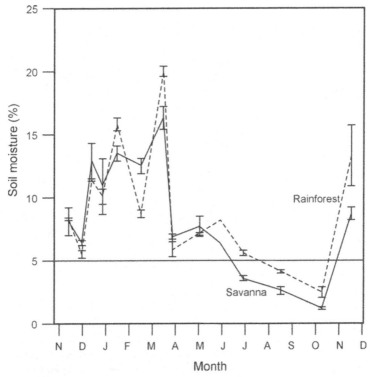

Figure 6.14
Variation in surface soil moisture content of monsoon rainforest and savanna vegetation. The standard errors are shown. The horizontal line is the wilting point (1.5 MPa) for similar soils as determined by Fensham and Kirkpatrick (1992). Adapted from Bowman and Panton (1993b).

ber), but reached similar levels to those recorded in the *E. nitens* forest by the late summer (February). Soils at 13 cm depth never reached wilting point, but *E. obliqua* forest had consistently lower levels of sub-surface soil moisture than the *E. nitens* and *Nothofagus cunninghamii* communities. The latter two communities had similar sub-surface moisture contents at the beginning of the summer but the rainforest had higher sub-surface moisture content at the end of the summer.

Rainforest water relations

A conspicuous feature of ecological research in Australian rainforests is the relative neglect of soil–plant water relationships. The few published studies show that many rainforest trees do not conform to the archetypal view of moisture dependent plants. A good example of a moisture-limited rainforest tree is the

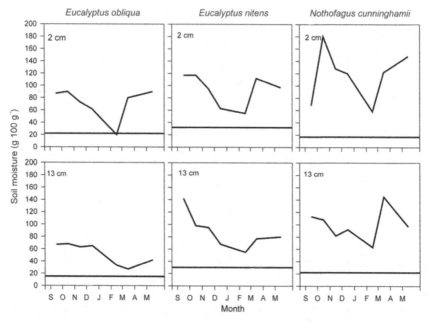

Figure 6.15

Wilting point (horizontal line) and moisture content of surface (2 cm) and sub-surface (13 cm) soils for Eucalyptus obliqua *forest,* Eucalyptus nitens *forest and a temperate* Nothofagus cunninghamii *rainforest. The soils were sampled over an 8-month period spanning the summer in 1967–1968. Adapted from Howard (1973b).*

humid tropical rainforest tree *Argyrodendron peralatum* (Doley *et al.* 1987, 1988). By noon on the third rainless day after a 23 mm rainfall event, the upper branches of a 32 m *Argyrodendron peralatum* tree were unable to control gas exchange and commenced to droop, switching from net photosynthesis to net respiration (Figure 6.16). Doley *et al.* concluded that 'some evergreen northern Australian rainforest species may spend at least part of the dry season in a condition of very limited carbon gain, or even of carbon loss'. Clearly species with such a physiological response would be incapable of surviving protracted periods of drought stress. Physiological studies have also shown the tolerance of some subtropical rainforest tree species to drought stress. Yates *et al.* (1988) showed that when dry season rainfall (May through to August) was less that 100 mm, *Flindersia collina, Excoecaria dallachyana* and *Mallotus philippensis* recorded shoot water potentials below −4 MPa (Figure 6.17). The rainforest conifer *Araucaria cunninghamii* showed even greater drought tolerance than these three broad-leaf species, having shoot water potentials that were unaffected by seasonal drought (> −2 MPa). The limited daily variation of shoot water

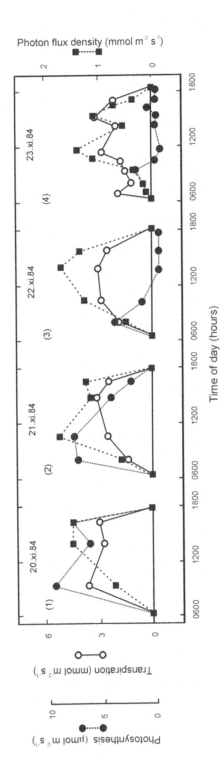

Figure 6.16
Variation in photon flux density, transpiration and assimilation rates for leaves on five branches at the top of a 32 m tall Argyodendron peralatum tree over four rain-free days following a thunderstorm that precipitated 23 mm of rain on 19 November 1984. Adapted from Doley et al. (1987).

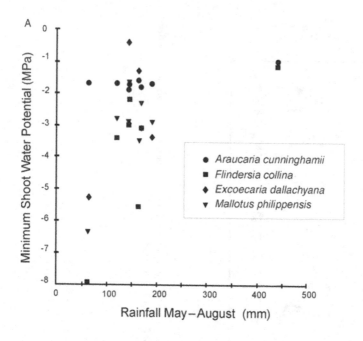

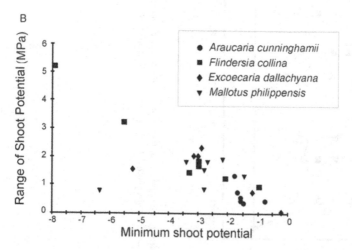

Figure 6.17

*Relationship between (A) dry season rainfall and minimum shoot water potential, and (B) minimum shoot water potential and the diurnal range of shoot water potential, for three broad-leaf rainforest species (*Flindersia collina, Excoecaria dallachyana *and* Mallotus philippensis*) and the rainforest conifer* Araucaria cunninghamii. *Adapted from Yates* et al. *(1988).*

TABLE 6.4. Estimated average drought damage of the crowns of mature rainforest trees in a gully in eastern Victoria in February 1983 following a severe drought

Species	% of foliage showing drought damage
Acacia mearnsii	12.5
Acmena smithii	17.0
Acronychia oblongifolia	9.5
Kunzea ericoides	6.0
Pittosporum undulatum	9.5
Rapanea howittiana	35.5
Tristaniopsis laurina	29.5

Adapted from Melick (1990a).

TABLE 6.5. Dawn water potentials of two Victorian rainforest trees measured in February 1983 on a north-facing slope in Woolshed Creek, eastern Victoria

Species	Dawn water potential (MPa)		
	Gully floor	Mid-slope	Top of slope
Acmena smithii	− 1.13	− 1.95	− 2.22
Pittosporum undulatum	N.A.	− 1.76	− 2.32

Adapted from Melick (1990a).

potentials in *A. cunninghamii* indicated a remarkable ability to control water loss under a range of environmental conditions.

Melick (1990a) showed that some Victorian rainforest gully species such as *Tristaniopsis laurina* and *Rapanea howittiana* suffered serious damage during drought (Table 6.4). Clearly, these species would not be able to survive on the drier, more drought prone soils on the slopes above the gully. However, some other rainforest species appear physiologically capable of surviving on drier and more exposed sites. Although *Pittosporum undulatum* and *Acmena smithii* suffered greater moisture stress on more exposed slopes, the level of stress was limited even at the top of a north-facing slope (Table 6.5). This finding is reinforced by the detailed ecophysiological studies of the tolerance of *Pittosporum undulatum* seedlings to water stress (Gleadow and Rowan 1982). They found that seedlings were able to recover from relative leaf water contents of less than 30% and xylem water potentials of less than -8.0 MPa, and that seedling drought resistance increased when they are grown in cool, shaded conditions similar to that found beneath dense *Eucalyptus* understoreys (Figure 6.18). This remarkable drought tolerance enables this species to invade unburnt dry *Eucalyptus* forests in central Victoria (Gleadow and Ashton 1981).

Unlike rainforest elsewhere in Australia, a large proportion of the rainforest tree species in hot, seasonally dry ('monsoonal') climates are deciduous, a habit which may assist them in avoiding water stress. Unwin and Kriedemann (1990) showed that leaf water potentials of the deciduous tree *Melia azederach* never fell below -2.0 MPa during the dry season (Table 6.1). However deciduousness varies considerably among different types of monsoon forest. Russell-Smith (1991) found that, in the 16 types of monsoon rainforest he recognised in the Northern Territory, the percentage of deciduous species varied from 16% to 56%. The correlation between the proportion of deciduous trees and mean annual rainfall estimated for the 16 types was low ($r = -0.34$), in part reflecting the ameliorating effect of perennial water supplies for some wet monsoon forest types and in part relating to soil fertility. Infertile soils typically support evergreen trees because of the high nutrient 'cost' associated with producing leaves (Webb 1968). Nonetheless, evergreen species are subjected to greater physiological stress than deciduous species in moisture limited climates. Unwin and Kriedemann showed that the evergreen species they studied had lower growth in the water-limited environment than did the deciduous species, and this pattern was reversed in the high rainfall environment (Table 6.1).

There is evidence that evergreen trees of savanna and monsoon rainforest are equally tolerant of seasonal drought. Bowman *et al.* (1999 *b*) compared rates of damage to foliage in evergreen tree seedlings of monsoon rainforest and long-unburnt savanna at a site in northeastern Queensland that had experienced three consecutive years in which rainfall had been less than half the 650 mm

---

Figure 6.18 (*opposite*)
Variation in (A) relative moisture content of leaves and (B) leaf water potential in Pittosporum undulatum *seedlings following cessation of watering. The seedlings were grown in cabinets with two contrasting treatments: (i) high temperature (27 °C 12 h day and 24 °C 12 h night), high light (128 µmol m^{-2} s^{-1}) and (ii) low temperature (21 °C 12 h day and 18 °C 12 h night) with low light (70 µmol m^{-2} s^{-1}). The first arrow indicates the time when wilting was first recorded and the second arrow indicates the time when all seedlings had wilted. Adapted from Gleadow and Rowan (1982).*

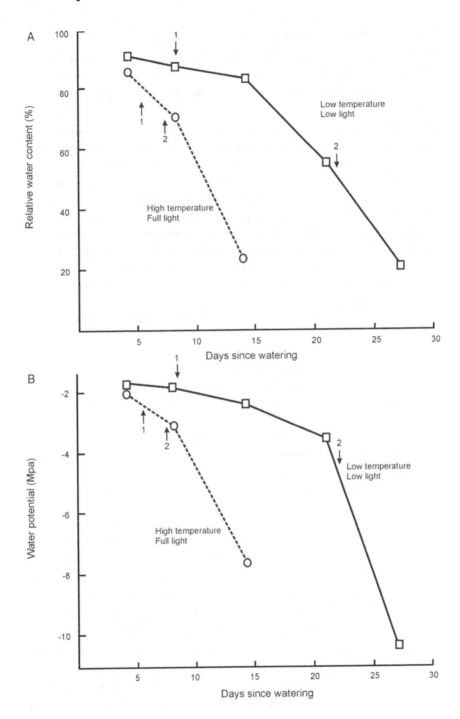

TABLE 6.6. Percentage of stems of evergreen rainforest and savanna species on a 1.56-ha grid at Pyro Pocket in north Queensland that had less than 40% of their foliage damaged by drought, in three stem size classes. The number of stems in each of the categories is shown in parentheses

For both the juvenile and sapling size classes, chi-square analysis revealed no significant difference in the frequency of undamaged evergreen rainforest and evergreen savanna stems. However, in the case of the saplings significantly ($P < 0.001$) more savanna species had damaged canopies compared with the rainforest species. The rainforest species were: *Alectryon connatus, Alectryon oleifolium, Alphitonia excelsa, Bursaria incana, Canthium oleifolium, Cupaniopsis anacardioides, Denhamia oleaster, Diospyros humilis, Drypetes lasiogyna, Exocarpos latifolius, Geijera salicifolia, Notelaea microcarpa* and *Polyscias elegans*. The savanna species were: *Dodonaea stenophylla, Dodonaea viscosa, Eucalyptus erythrophloia, Eucalyptus xanthocalda, Hakea lorea* and *Maytenus cunninghamii*.

Species type	% undamaged (*n*) Size class		
	Juvenile	Sapling	Tree
Evergreen–savanna	66 (374)	48 (133)	11 (156)
Evergreen–rainforest	64 (109)	71 (56)	0 (0)

Adapted from Bowman *et al.* (1999*b*).

annual average, and found no significant difference (Table 6.6). Furthermore, significantly ($P < 0.001$) more savanna saplings had damaged canopies compared to evergreen rainforest species. Ecophysiological studies demonstrate the drought tolerance of the evergreen monsoon rainforest tree *Allosyncarpia ternata*, a tree that dominates rainforest on the western escarpment of the Arnhem Land Plateau in the Northern Territory (Fordyce *et al.* 1997*a*). They found that in a permanently moist ravine, the trees showed no evidence of moisture stress and that although trees on a hill-top suffered some moisture stress in the dry season (Figure 6.19) they were able to continue to transpire and photosynthesise (Figure 6.20). Fordyce *et al.* assumed that *Allosyncarpia* trees were able to extract moisture reserves held in porous bedrock and deep siliceous soils. Working in

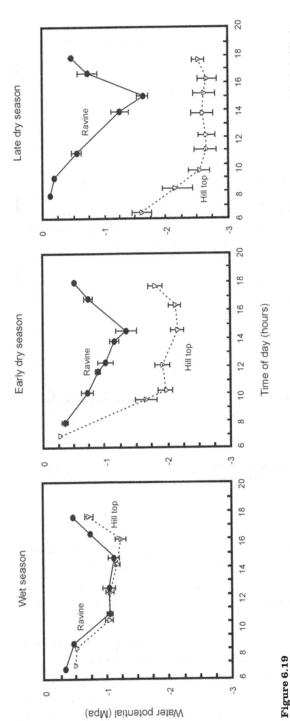

Figure 6.19

Seasonal and diurnal variation in the leaf water potential of Allosyncarpia ternata trees growing in a moist ravine (solid line) and on an adjacent dry hill (dashed line). The bars are the standard errors of the means. Adapted from Fordyce et al. (1997a).

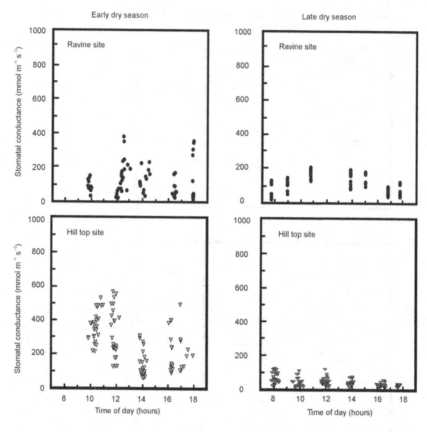

Figure 6.20

Early and late dry season and diurnal variation in stomatal conductance of Allosyncarpia
ternata *trees growing in a moist ravine (Site 1) and on an adjacent dry hill (Site 2). Adapted
from Fordyce* et al. *(1997a).*

tropical savannas, Myers *et al.* (1997) have also suggested that available ground
water explains the low levels of moisture stress observed in evergreen *Eucalyptus
tetrodonta* trees. These studies suggest that there is no physiological reason why
monsoon rainforest trees such as *Allosyncarpia ternata* could not grow in *Eucalyptus* savannas.

Barrett *et al.* (1996) identified the importance of sub-surface moisture
supplies for a study area in southeastern New South Wales in which *Ceratopetalum apetalum–Doryphora sassafras* rainforest grew on the lower slopes and
floor of a gully, with *Eucalyptus maculata* forest on the surrounding slopes. On
both the north-facing and south-facing slopes, the moisture content of the

rainforest soils was higher than in the adjacent *Eucalyptus* forest soils, but the moisture content of the *Eucalyptus* forest soils on the south-facing slope was higher than in the rainforest soils on the north-facing slope (Figures 6.21 and 6.22), a difference they attributed to evapotranspiration. Estimates of water use did not mirror the availability of surface soil moisture. Surface soil moisture declined during the summer, as did the moisture content of sapwood in the rainforest trees (Figure 6.22). But there was no decline in the moisture content of *Eucalyptus* sapwood. It was assumed that the *Eucalyptus* trees obtained moisture from the sub-soil through the summer, an assumption that is reinforced by estimates of total daily water use. Although the rainforest canopy had the larger leaf area, the *Eucalyptus* forest had the greatest daily water use because eucalypts have larger vessels and higher sap flow rates (Figure 6.23). Barrett *et al.* suggested that high water tables at the foot of slopes during winter precluded the development of deep root systems in rainforest trees, exposing the rainforest trees to moisture stress in the summer. It is unclear whether rainforest trees would be able to obtain ground water on slopes in the absence of restrictive winter water tables. The restriction of rainforest to the lower parts of the south-facing slope is not explicable by water stress alone because surface soil moisture higher up the slope was greater than in rainforest soils on the north-facing slope. Barrett and Ash (1992) explained the restriction of rainforest to gully floors as being related, in part, to past fires.

A number of authors have argued that the patchy occurrence of monsoon rainforest simply reflects the localised occurrence of underground supplies of moisture (Beard 1976; Hnatiuk and Kenneally 1981; Braithwaite *et al.* 1984; Bowman and Wightman 1985). This view has been most forcefully enunciated by Specht (1981*b*), who used ecophysiological principles in a theory that sought to explain the occurrence of rainforest throughout Australia. According to Specht's theory, rainforest ('closed forests' in Specht's terminology) can only occur where there are close to optimal soil moisture supplies throughout the year. This theory is also used by Specht (1981*a*) to bolster his classificatory system of Australian vegetation. Given the widespread usage of Specht's classificatory system, it is remarkable how little critical discussion there has been of his theory, especially the basic assumption that mature vegetation is in equilibrium with climate and the specific prediction that natural isolates of rainforest are dependent upon perennial moisture supplies. I therefore provide a brief sketch and critique of Specht's theory.

Specht (1972, 1981*a*) applied a moisture balance model he had previously developed for *Eucalyptus* and heath communities (Martin and Specht 1962;

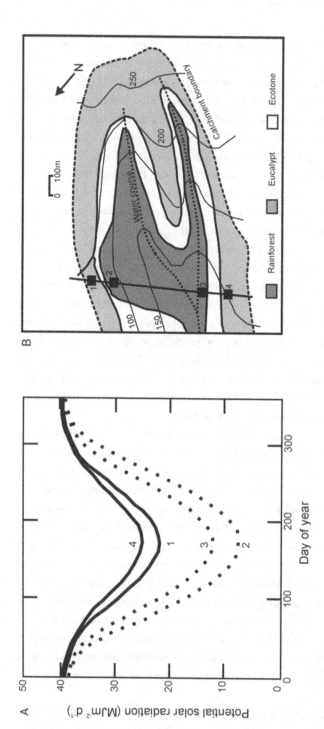

Figure 6.21

(A) Estimated daily solar radiation received at each of the four sites where: 1: south-facing top-slope; 2: south-facing foot-slope; 3: north-facing foot-slope; and 4: north-facing top-slope. (B) Vegetation map and location of the four sites used by Barrett et al. (1996) in their water use study. Adapted from Barrett et al. (1996).

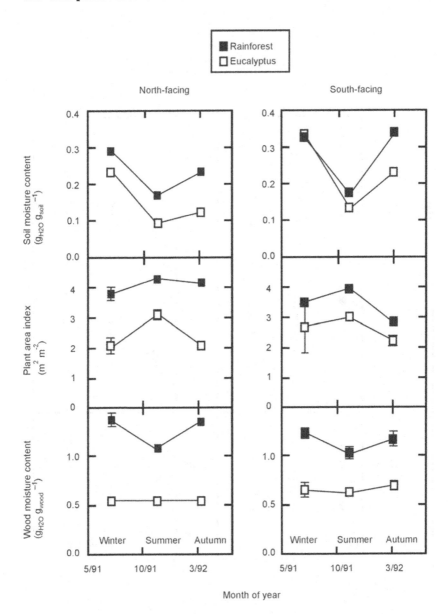

Figure 6.22

Seasonal variation in surface soil moisture content, plant area index, and sapwood moisture content for rainforest and Eucalyptus *forest on north- and south-facing slopes of a small catchment in south-eastern NSW. Adapted from Barrett* et al. *(1996).*

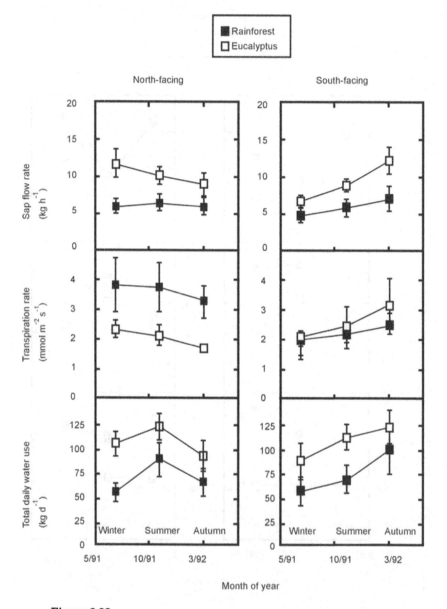

Figure 6.23

Seasonal variation in average maximum sap flow rate, average maximum transpiration rate and average daily water use for rainforest and Eucalyptus forest on north- and south-facing slopes of a small catchment in south-eastern NSW. Data are averages from eight trees in four adjacent plots. Adapted from Barrett et al. (1996).

Specht 1957, 1972, 1981b; Specht and Jones 1971) to explain the occurrence of closed-canopy forests in Australia. The model is a linear equation:

$$E_a = kWE_o \text{ or} \qquad \qquad (1a)$$
$$E_a/E_o = kW \qquad \qquad (1b)$$

where E_a = actual evapotranspiration (cm of water)
        E_o = pan evaporation (cm of water)
        W = water extractable from the soil by vegetation (cm of water).

The term W is derived from the sum of precipitation (P) (less run-off and drainage) and extractable soil moisture carried over from the previous month ($S_{ext(n-1)}$), i.e.:

$$W_n = P_n + S_{ext(n-1)}. \qquad \qquad (2)$$

Run-off and drainage (D) is assumed to be zero unless the soil moisture exceeds the hypothesised maximum value of S_{max}, the maximum storage capacity of the soil, i.e.:

$$S_{ext(n)} > S_{max} \text{ then } D_{(n-1)} = S_{ext(n)} - S_{max} \qquad \qquad (3)$$

The constant k is thought to be an integrated measure of the evaporative capacity of the canopy. For a given level of pan evaporation (E_o), k represents the rate of change between evapotranspiration (E_a) with respect to extractable soil moisture (W). The constant k is thus the slope of the linear relationship between relative evapotranspiration (E_a/E_o) and extractable soil moisture (W). The model assumes that all precipitation enters the soil, that the terrain is level and that the plant community is evergreen with a relatively constant canopy area. The model would have to be substantially modified to take into account the effect of seasonal variation in foliage projective cover, and variation in aspect, topography, soil water penetration and root distribution (Specht 1972). The study of Barrett et al. (1996) clearly demonstrated that slope aspect confounded a simple relationship between whole-crown transpiration and leaf area (Figures 6.22 and 6.23). The simultaneous measurement of E_o, P and S_{ext} necessary to derive values of E_a and k requires detailed field studies, and there have been only a few such studies conducted in Australia (Specht 1957; Martin and Specht 1962; Specht and Jones 1971; Pressland 1976). Specht (1972, 1981b) provided a brief outline of how the terms E_a, k, and W can be estimated through iteration in a computer program given only E_o and P as inputs. The model determines the maximal value of k in equation (1) such that W in all months is > 0.

In essence, Specht assumed that climate determines the physiognomy of mature vegetation (Specht 1981a). He believed that climatic equilibrium in Australian vegetation has arisen by 'evolutionary adjustment' such that a given plant community will use the maximum amount of water possible (i.e. maximum k) that will allow it to survive the driest period of the year (W always > 0) (Specht 1972). The notion that vegetation types 'evolve' is not elaborated by Specht but clearly there are theoretical difficulties concerning selective pressure on individual plants versus groups of plants. Perhaps Specht's use of the term 'evolution' signals his belief that Australian vegetation types are highly integrated systems of great antiquity. Clearly, infrequent severe droughts are of more evolutionary importance in determining k than the driest months in 'average' years. It must be recognised that estimated values of k will increase with the length of the climatic data set used to calculate it. Specht (1972) used climatic averages derived from a 30-year sampling period, which is unlikely to be sufficient to incorporate the climatic extremes encountered in an area.

Significant linear relationships between computer estimates of optimal values of k and the measured foliage projective covers (FPC) of understorey and overstorey strata from 'mature' Australian vegetation were interpreted by Specht (1983) as empirical support for his model (Figure 6.24). This interpretation is dependent on the unrealistic assumption that vegetation structure is static. As will be discussed in subsequent chapters, long fire-free periods allow rainforest to invade forest, thereby changing canopy cover.

Field studies do not support the predictions of the model. Specht *et al.* (1977) argued that monsoon rainforests near Weipa in the Australian monsoon tropics are restricted to sites where roots can tap aquifers, and that in the absence of aquifers these rainforests could only colonise the *Eucalyptus* savannas if the amount of precipitation retained in the soil were increased to 'almost 120%' of current values. However, Bowman and Fensham (1991) reported that extensive geological drilling and subsequent strip mining showed that a monsoon rainforest in the Weipa area was not situated above a shallow (< 5 m) aquifer. Langkamp *et al.* (1981) also rejected the idea that monsoon rainforests were restricted to sites above shallow aquifers.

The predictions of Specht's model were not supported by Pressland's (1976) experimental studies. He determined k for three replicate stands of *Acacia aneura*, an arid zone tall shrub, that were thinned to five different densities. He calculated k from linear regressions of observed relative evapotranspiration plotted against soil moisture (W) that he measured to a depth of 135 cm over a period of 2 years. Figure 6.25 shows that there is no relationship between

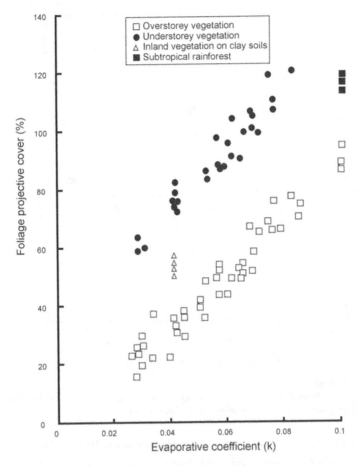

Figure 6.24

Relationship between the foliage projective cover (FPC) and evaporative coefficient (k) for the overstorey (open squares) and understorey (closed circles) of 56 Australian plant communities. Closed squares denote subtropical rainforest communities including overstorey and ground cover but not epiphytes. Triangles are for plant communities on heavy clay soils in central Queensland. FPC is defined by Specht (1983) as 'the percentage of land covered by photosynthetic organs, observed vertically' and therefore is not the same as the total area of foliage. The vegetation Specht (1983) sampled was assumed to have 'reached a steady-state relationship for the FPCs of both overstorey and understorey strata'. Adapted from Specht (1983).

estimated values of k and *Acacia* stem density. Pressland noted that all k values of between 0.04 to 0.08 he determined for *Acacia aneura* shrubland were 'higher than those suggested by Specht for a mulga [*Acacia aneura*] woodland'. Indeed, the k values for *Eucalyptus obliqua* open forest (0.053) and *E. goniocalyx* open forest (0.046) in the Mount Lofty Ranges in South Australia (Specht 1972) are

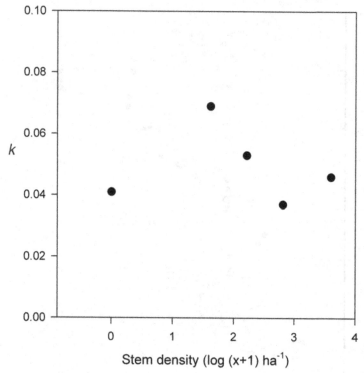

Figure 6.25

Relationship between evaporative coefficient (k) and stem density of thinned Acacia aneura *shrublands in southwestern Queensland. Derived from Pressland (1976).*

within the same range of *k* values reported by Pressland (1976). The juxtaposi-tion of closed-canopy *Acacia harpophylla* communities with open-canopied *Eucalyptus populnea* communities was explained by Specht (1981*c*) as being a consequence of compensating differences in leaf resistance of these plant com-munities. Thus, physiological differences may account for the similarity between the *k* values of *Eucalyptus* forests and *Acacia aneura* shrublands measured by Pressland. The physiological capacity of many rainforest plants to tolerate seasonal drought makes Specht's assumption that evapotranspiration from closed canopy forests is always maximal questionable (Specht 1981*a*). To some degree Specht accepted this by noting that drought stress may produce semi-deciduous rainforests in climates that have dry seasons with high evaporation levels. Nonetheless evergreen rainforest often occurs on sites subjected to severe seasonal moisture deficits.

Conclusion

There is a mass of evidence to reject the view that Australian rainforests are axiomatically restricted to sites that are permanently drought-free. Some rainforests occur in seasonally arid climates without supplementary water supplies such as localised aquifers. Ecophysiological studies demonstrate that tree species from a wide cross section of Australian rainforest have considerable endurance of low xylem pressure potentials. At the local scale, rainforest boundaries are not necessarily correlated with gradients in soil moisture availability. While it is true that variation in moisture supply influences species composition and structure of rainforest vegetation, even this relationship is complicated by the effects of soils, topography and fire history. The wide tolerance of some rainforest trees to drought places a question mark over the importance of aridification in influencing the distribution of rainforest over geological time. Clayton-Greene and Beard (1985) realised that the hypothesis that rainforest is moisture dependent is incompatible with the biogeography of rainforest in the Kimberley because the fragmentary distribution of these rainforests could not have developed since the end of the Pleistocene aridity 10 000 years ago. Instead, they explained the distribution of rainforest as being a consequence of landscape fire.

7

The climate theory II. Light and temperature

Introduction

An obvious feature of the boundary between rainforests and adjacent non-rainforest formations is a dramatic change in microclimate. Seddon (1984) suggested that microclimate is often unconsciously used to dichotomise rainforest from non-rainforest in Australia. He wrote that rainforests are 'the only forest form that is at all common in Australia that *does* cast a dense shade' being 'so different from the familiar, light-drenched open forest dominated by eucalypts' (original emphasis) (Figure 7.1). It has often been assumed that the differences in microclimate are of significance in controlling the establishment of seedlings. Herbert (1932) wrote that 'in the struggle for existence in the crowded rainforest, sun-loving types, such as *Eucalyptus*, have no chance of becoming established'. Another factor that changes across the rainforest boundary is ground surface temperature. Minimum temperature and frost are also thought to prohibit subtropical and tropical rainforest from occurring on otherwise optimal sites (Webb and Tracey 1981) and past colder climates may account for the occurrence of grasslands within tracts of subtropical rainforest (Webb 1964). The purpose of this chapter is to:

  (i)   describe the difference in light across rainforest boundaries;

  (ii) evaluate the hypothesis that differences in light regime control regeneration of rainforest and non-rainforest trees;

  (iii) consider the effect of maximum and minimum temperature on rainforest distribution.

Figure 7.1
A distinguishing feature of Eucalyptus *forests is the openness of the canopy, as illustrated by this eastern Victorian forest. (Photograph: Don Franklin.)*

Light environment across the rainforest boundary

There have been few studies that have quantified the change in light regime across the rainforest boundary in Australia. Turton and Duff (1992) used hemispherical photographs taken 1.3 m above the ground surface to demonstrate that there is a steep gradient in diffuse and direct photosynthetic active radiation (PAR) across an east-facing rainforest boundary on nearly level terrain in the Australian humid tropics (Figure 7.2). The *Eucalyptus* forest was estimated to absorb 55% of the average direct and diffuse PAR, while the rainforest absorbed almost all the PAR. Turton and Duff (1992) found considerable seasonal variation in direct, but not diffuse, PAR in the *Eucalyptus* forest, but little seasonal variation in the rainforest. Subsequent research has revealed that the nature of the light gradient across rainforest boundaries varies locally. Also working in the humid tropics, Turton and Sexton (1996) showed that there were significant linear relationships between distance from the rainforest interior and March and June PAR for four transects across a rainforest boundary. However, there was also substantial scatter in these relationships (Figure 7.3), and the slopes of the regressions were significantly different. The cause of the variation in the light gradient across the rainforest boundaries is not understood, although Turton and Sexton (1996) suggested that slope angle, aspect and different stages

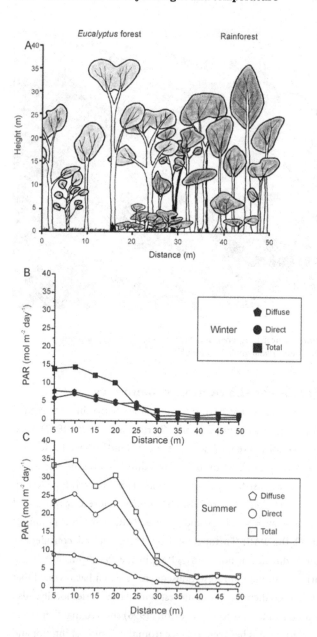

Figure 7.2

Profile of a Eucalyptus-rainforest boundary across a 3° slope on the Atherton Tablelands in north Queensland (A), and variation in photosynthetically active solar radiation across it (B and C). The direct, diffuse and total potential (i.e. under cloud-free skies) photosynthetically active radiation (PAR) is shown for both the winter solstice (B) and summer solstice (C). The solar radiation was expressed as mols of photosynthetic photons (400–700 nm) per m² per day. Adapted from Turton and Duff (1992).

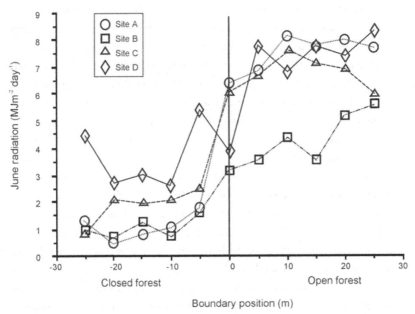

Figure 7.3

Variation in the estimated June daily solar radiation for four rainforest boundaries on the Atherton Tablelands. Almost identical trends were detected for estimated March radiation levels. The topographic positions of the four sites were as following: site A was on a ridgeline where the rainforest occurred on an 8–19° slope with a southerly aspect while the Eucalyptus *forest occurred on the opposite north-facing slope with an angle of 0–14°; at site B, both the rainforest and* Eucalyptus *forest were on a 19° slope with a southeasterly aspect; at site C the rainforest was on a 13° slope with a northwesterly aspect, while the* Eucalyptus *forest was up-slope on both the ridge-top and on the opposite southeastern slope of 12°; at site D, both the rainforest and the* Eucalyptus *forest occurred on 10° slope with an easterly aspect. Adapted from Turton and Sexton (1996).*

in a fire-controlled rainforest succession may be responsible. Light quality must also vary across rainforest boundaries although the exact nature of these differences awaits further inquiry. Within *Atherosperma moschatum* rainforest, Olesen (1992) demonstrated that the proportion of blue, green, red and near infra-red light was influenced by canopy absorption, reflection and transmission characteristics. Clearly these latter traits vary across rainforest boundaries.

The generalisation that the floor of non-rainforest formations receives high PAR only holds if there is a sparse understorey. A feature of many *Eucalyptus* forests in high rainfall environments is the development of a dense shrub layer that casts a deep shade on the forest floor. For example, Ashton and Turner (1979) found that a *Pomaderris aspera* understorey beneath a *Eucalyptus regnans* forest substantially reduced solar radiation in both summer and winter. Beneath the *Eucalyptus regnans* canopy, the summer and winter solar radiation was 61% and 51% of clear sky light levels respectively. The corresponding measurements beneath the *Pomaderris aspera* understorey were only 19% and 16% of clear sky levels. Further, while the generalisation that rainforests are light-limiting environments is true, it must be recognised that there is enormous spatial and temporal variation. Bjorkman and Ludlow (1972) showed that on a clear autumn day the forest floor under dense subtropical rainforest in southeast Queensland received 0.4% of above-canopy PAR. Sun-flecks contributed 61% of this tiny quantity of light. Osunkjoya *et al.* (1992) found that ground surface PAR within tree-fall gaps in a humid tropical rainforest on the Atherton Tablelands received 4–9% of the above-canopy PAR, while the forest floor beneath the intact canopy received 0.5–2.5%. Working on the Atherton Tablelands, Turton (1990) found that during the winter solstice an undisturbed montane rainforest canopy reduced incoming PAR to 1–4% while the floor in a small gap (150 m^2) received 10% of the incoming PAR.

There are no studies of how light regime varies in response to seasonal leaf flushing and leaf fall. Although Lowman (1986) found that there was no significant seasonal variation in the amount of visible sky in three different evergreen rainforest types (Figure 7.4), this may not be the case for rainforests dominated by deciduous species. Bowman and Wilson (1988) found that the mean canopy cover of dry monsoon forest changed from 74% at the end of a dry season to 84% at the end of the following wet season. Periodic defoliation following severe storms has been shown to substantially change light climate on the forest floor. Turton (1992) found that following a tropical cyclone, solar radiation on the forest floor increased from 2.5–3.4% to 6.0–8.6% of clear sky levels and that the spatial variation of the light regime was substantially reduced (Figure 7.5). However, these effects were short-lived, with the canopy recovering within 12 months.

Seedling growth response to light

Many ecologists have held the belief that rainforest tree seedlings are highly shade tolerant while *Eucalyptus* and other non-rainforest tree species are light demanding (Herbert 1932; Fraser and Vickery 1939; Francis 1951; Jackson 1968; Unwin 1989). This view is not substantiated by research. Rainforest tree seedlings are tolerant of a wide range of light conditions, some species are highly light demanding and very few are truly shade tolerant. Conversely, some non-rainforest tree seedlings are surprisingly shade tolerant. A brief review of studies of the shade tolerance of rainforest and non-rainforest species is presented.

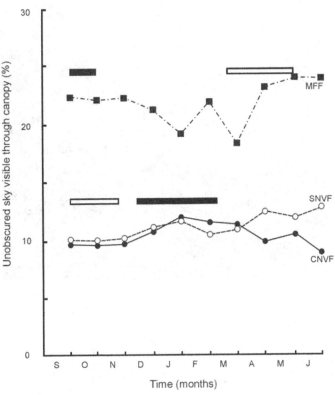

Figure 7.4

Seasonal variation in the canopy coverage of three types of rainforest (cool temperate or MFF, warm temperate or SNVF, and subtropical or CNVF) in central New South Wales. The percentage of unobstructed sky visible through the canopies was measured with hemispherical photographs. Open bars indicate periods of leaf fall and solid bars indicate periods of leaf flushing. Adapted from Lowman (1986).

Ecological studies of *Nothofagus gunnii*, *Podocarpus lawrencei* and *Athrotaxis selaginoides* suggested that these temperate rainforest species could not regenerate under dense evergreen canopies (Read and Hill 1985; Cullen 1987; P. Barker 1991). Indeed, the growth cabinet experiment with seedlings of six Tasmania rainforest tree species undertaken by Read (1985) demonstrated that all had maximal growth under conditions of high light. Under low light conditions, only *Atherosperma moschatum* was able to maintain a high relative growth rate, *Eucryphia lucida* and *Phyllocladus aspleniifolius* had negative growth rates, *Lagarostrobos franklinii* had zero growth and *Nothofagus cunninghamii* and *Athrotaxis selaginoides* had very low growth rates (Figure 7.6). Barrett and Ash (1992) found that light, as opposed to nutrient or moisture supply, most strongly controlled seeding growth rate of all seven rainforest and non-rainforest species

Figure 7.5
Defoliation of a humid tropical rainforest following a severe cyclone. (© Murray Fagg, Australian National Botanical Gardens.)

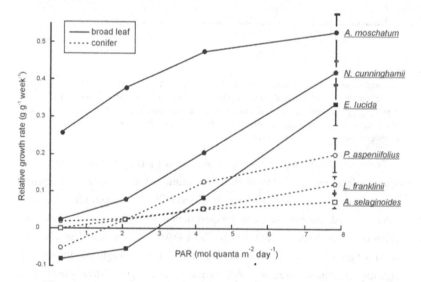

Figure 7.6
*The relative growth rates of three Tasmanian broad-leaf species (*Nothofagus cunninghamii, Atherosperma moschatum *and* Eucryphia lucida*) and three Tasmanian conifers (*Athrotaxis selaginoides, Lagarostrobos franklinii *and* Phyllocladus aspleniifolius*). Five to ten seedlings of each species were grown for 35 days under four light treatments in a growth cabinet with a photoperiod of 12 hours. Day temperature was set at 23 °C and night temperature at 20 °C. The largest standard error is shown for each species. Adapted from Read (1985).*

they tested. They also found all species tested had maximum growth rates under high light conditions and that there was negligible difference in growth rates in the shade (Figure 7.7). These data are inconsistent with Barrett and Ash's (1992) conclusion 'that physiological responses to environmental gradients may determine much of the distribution of rainforest and eucalypt vegetation in southern New South Wales'.

Field and shade-house studies showed that regardless of differences in seed size and successional status, the seedling growth of six humid tropical rainforest canopy trees was positively correlated with PAR (Figure 7.8) (Osunkoya *et al.* 1993). Indeed, maximum biomass production of potted seedlings occurred at the highest level tested (37% full sun PAR) (Osunkoya and Ash 1991). In heavy shade (< 2% full sun PAR), there was high mortality of seedlings (Osunkjoya *et al.* 1992). Osunkoya *et al.* (1993) concluded that none of the species they tested could be considered truly shade tolerant, and optimal growth of *Polyscias elegans* and *Flindersia brayleyana* occurred under high light. Growth chamber experiments showed that seedlings of *Flindersia brayleyana* were tolerant of a wide range of light conditions. Further, growth of this humid tropical rainforest tree was more strongly influenced by light regime than nutrient supply, because under low light conditions seedlings were unresponsive to high levels of nutrients while maximum growth occurred under high light and nutrient conditions (Thompson *et al.* 1988).

A limitation to seedling studies is that they do not consider possible changes in photosynthesis associated with plant development. For example, juveniles of some temperate *Eucalyptus* species are more tolerant of shade than older plants (Cameron 1970; Ashton and Turner 1979). Admittedly no such ontogenetical shifts in the photosynthetic performance of rainforest species are known, but there are shifts in the dominance patterns of rainforest seedlings in tree-fall gaps. For instance, Thompson *et al.* (1988) contrasted the growth of *Acacia aulacocarpa*, *Toona australis*, *Flindersia brayleyana* and *Darlingia darlingiana* grown from seeds in tree-fall gaps that received 40, 9 and 0.6% full sunlight respectively. They showed that under 40% sunlight, *Acacia aulacocarpia* was dominant for the first 4 years but by 7 years of seedling growth it was overtaken by *Toona australis*. In the gap that received 9% of full sunlight, *Toona australis* was the dominant at the 4-year measurement but after 7 years it was subordinate to *Darlingia darlingiana* and *Flindersia brayleyana*. *Flindersia brayleyana* persisted for 4 years in a gap that received 0.6% sunlight but after 7 years the only surviving species in this gap was *Darlingia darlingiana*. This highlights the need for long-term studies to understand the growth and competition of trees under varying light regimes.

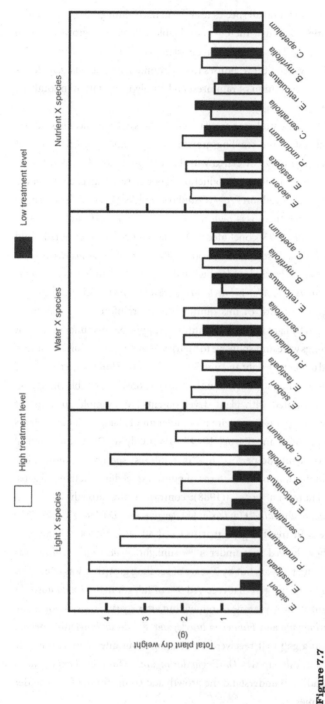

Figure 7.7

Variation in mean total plant biomass for seven tree species grown with high and low levels of light, nutrients and moisture supply. These species are thought to be characteristic of Eucalyptus forest (E. fastigata and E. sieberi), ecotone (Callicoma serratifolia, Pittosporum undulatum and Elaeocarpus reticulatus) and rainforest (Ceratopetalum apetalum and Backhousia myrtifolia). They were grown in a glasshouse for 22 weeks under high and low light (1230–1670 vs 200–530 μmol m⁻² s⁻¹), high and low nutrients, and high and low moisture conditions. Adapted from Barrett and Ash (1992).

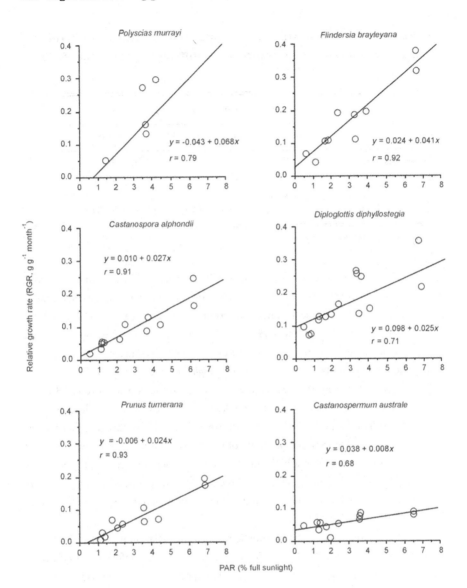

Figure 7.8

Relationship between photosynthetically active radiation and relative seedling growth rates of six species of rainforest tree. Light levels below 2.5% reflect forest interior environments and above 2.5% correspond to gaps of different sizes. Adapted from Osunkoya et al. (1993).

TABLE 7.1. The biomass (g) and percentage survival (in brackets) of four non-rainforest species grown under different light regimes for 9 months

Species	Relative irradiance (% of sunlight)			
	4	8	30	100
Eucalyptus ovata	0 (0)	3.97 (65)	16.28 (70)	34.47 (100)
Casuarina littoralis	2.32 (70)	7.39 (80)	16.44 (85)	15.64 (80)
Casuarina stricta	2.10 (65)	10.70 (85)	34.47 (80)	27.39 (85)
Acacia pycnantha	7.46 (90)	12.43 (85)	32.14 (90)	21.54 (90)

Adapted from Withers (1979).

Low light conditions can also have indirect effects on the growth and survival of some species. For example, shade can decrease the rate of mycorrhizal infection and increase the rate of infection by pathogens (Ashton 1976*b*; Ashton and Turner 1979; Withers 1979). This may explain why *Eucalyptus regnans* seedlings cannot survive on the forest floor despite having a growth compensation point similar to *Nothofagus cunninghamii* and *Pittosporum undulatum*. These latter rainforest species are able to invade *E. regnans* forests (Howard 1973*b*; Ashton and Turner 1979; Gleadow and Ashton 1981; Gleadow *et al.* 1983). Not all non-rainforest species are as intolerant of shaded conditions as *Eucalyptus*. For example, seedlings of *Casuarina littoralis*, *Acacia pycnantha* and *Casuarina stricta* had a high survival rate under 4% sunlight, and growth was maximal under 30% sunlight (Withers 1979). In contrast, *Eucalyptus ovata* seedlings died in 4% sunlight, and maximum growth of this species occurred in 100% sunlight (Table 7.1).

Measurements of photosynthetic response to variation in light reinforce the conclusion that there is considerable overlap between rainforest and non-rainforest species (Table 7.2). Some rainforest species (e.g. *Nothofagus gunnii*, *Tristaniopsis laurina* and *Agathis robusta*) have similar light saturated photosynthesis (LSP) values to the few *Eucalyptus* species that have been tested. Conversely, *Casuarina littoralis* has a lower LSP than many rainforest species. There is also great intra- and inter-specific variation in light compensation point (LCP). The two non-rainforest species for which data are available (*Casuarina littoralis*, *Eucalyptus ovata*) have LCPs lower than many rainforest species (Table 7.2). Light saturated assimilation rate (A_{max}) also shows considerable variation between rainforest species (Read 1985; Turnbull 1991). Turnbull concluded that

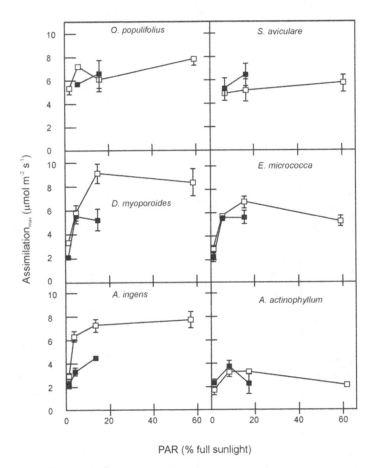

Figure 7.9

Variation in the light-saturated assimilation rate (A_{max}) relation to growth irradiance (% full sun PAR) for seedlings of six rainforest tree species from early- (Omalanthus populifolius, Solanum aviculare), *mid-* (Duboisia myoporoides, Euodia micrococca) *and late-* (Acmena ingens, Argyrodendron actinophyllum) *successional stages grown in a greenhouse environment. With the exception of* Solanum aviculare, *there was a significant decrease in light saturated assimilation (A_{max}) with decreasing growth irradiance (% full sun PAR). Open symbols correspond to unfiltered light and closed symbols correspond to filtered light with a red (655–665 nm) to far red (725–735 nm) ratio of 0.2. Of the six species tested, only* Acmena ingens *and* Duboisia myoporoides *had significantly ($P < 0.01$) lower A_{max} under filtered light. Bars indicate ± one standard error of the mean. Missing data points correspond to treatments where no individuals survived. Adapted from Turnbull (1991).*

successional status was not well correlated with seedling A_{max} of subtropical rainforest species. Seedlings of the primary rainforest tree species *Acmena ingens* were able to maintain high A_{max} under 60% full sun PAR. Early successional tree species such as *Omalanthus populifolius* and *Solanum aviculare* maintained high A_{max} under low light conditions (Figure 7.9). In that study, there was no

TABLE 7.2. Light saturation and compensation points for seedlings of a range of rainforest and non-rainforest tree species

Values are means (range). Seedlings were grown under a range of different light environments in greenhouses, shade-houses and growth cabinets

Forest types	Species	Light saturation point (μmol quanta m^{-2} s^{-1})	Light compensation point (μmol quanta m^{-2} s^{-1})
Temperate non-rainforest	Casuarina littoralis	500	5.7
	Eucalyptus ovata	> 1000	9.1
Temperate rainforest	Atherosperma moschatum	330 (250–760)	7.6 (4.2–21)
	Elaeocarpus holopetalus	> 550	5.8 (4.6–7.1)
	Nothofagus cunninghamii	600 (340–800)	19.5 (6–31)
	Phyllocladus aspleniifolius	610 (330–770)	13 (7–22)
	Lagarostrobos franklinii	760 (740–780)	14 (8–17)
	Eucryphia lucida	790 (130–1500)	18 (8–30)
	Athrotaxis selaginoides	1000 (740–1500)	13 (6–20)
	Nothofagus gunnii	1040	62
Temperate and subtropical rainforest	Ceratopetalum apetalum	430 (90–770)	8 (6–9)
	Eucryphia moorei	540 (340–740)	13 (11–14)
	Nothofagus moorei	565 (330–800)	18 (13–22)
	Doryphora sassafras	570 (340–800)	9 (6–12)
	Acmena smithii	1100	12 (7–16)
	Tristaniopsis laurina	1800 (1100–2500)	30 (20–39)

			N.A.
Subtropical and tropical non-rainforest	*Eucalyptus grandis*	1000 (800–1200)	N.A.
Subtropical and tropical rainforest	*Argyrodendron actinophyllum*	495 (270–720)	14 (5–22)
	Flindersia brayleyana	505 (340–710)	23 (17–37)
	Agathis microstachys	610 (390–830)	15 (5–24)
	Omalanthus populifolius	820 (670–970)	10 (3–17)
	Agathis robusta	1125 (750–1500)	19 (15–23)

N.A.: not available

Data compiled from Read (1985), Read and Hill (1985), Melick (1990*b*), Turnbull (1991), Langenheim *et al.* (1984), Withers (1979), Doley (1978), Thompson *et al.* (1988) and Olesen (1997).

TABLE 7.3. Comparison of photosynthetic and morphological attributes of *Allosyncarpia ternata* leaves from trees growing in a canyon and on an adjacent hill-top in Kakadu National Park

Values are mean (standard error)

Leaf parameter	Ravine floor	Hill-top
Morning A_{max} (μmol CO_2 quanta m^{-2} s^{-1})	5.9 (0.01)	7.9 (0.26)
Afternoon A_{max} (μmol CO_2 quanta m^{-2} s^{-1})	5.4 (0.31)	4.8 (0.24)
Morning apparent quantum yield	0.028 (0.004)	0.014 (0.005)
Afternoon apparent quantum yield	0.019 (0.003)	0.010 (0.002)
Light saturation point (μmol quanta m^{-2} s^{-1})	440	960
Light compensation point (μmol quanta m^{-2} s^{-1})	14	28
Leaf area (mm^2)	2544 (90)	1597 (72)
Specific leaf area (mm^2 g^{-1})	9207 (210)	7223 (158)
Chlorophyll *a* (mg g^{-1} leaf dry weight)	2.8 (0.26)	1.5 (0.42)
Chlorophyll *b* (mg g^{-1} leaf dry weight)	1.0 (0.10)	0.6 (0.11)

Adapted from Fordyce *et al.* (1995).

systematic response of seedlings from early or late successional stages to differences in photosynthetic light quality (neutral shade versus filtered shade with a low red to far-red ratio).

Not only do many Australian tropical rainforest species display a wide tolerance of light regimes, but there is also evidence that some tropical rainforest species can acclimatise to both decreases and increases in PAR. This is achieved by adjusting the photosynthetic performance of existing leaves or by replacing older leaves with suitably acclimatised new foliage (Langenheim *et al.* 1984; Pearcy 1987; Osunkoya and Ash 1991). Figure 7.10 illustrates that the relationship between assimilation and photon flux density varied between canopy and understorey plants (Pearcy 1987). He also found that the photosynthetic response curve for leaves of *Argyrodendron peralatum* on a stem in a forest gap environment with intermediate photon flux density was intermediate compared with that of the canopy and understorey. Physiological acclimatisation of leaves is often accompanied by substantial difference in leaf morphology and chemistry. Leaf specific weight, leaf nitrogen content and leaf chlorophyll content of *Argyrodendron peralatum* leaves all increased with increasing light associated with the gradient from the understorey to the forest canopy (Pearcy 1987). Similarly, Fordyce *et al.* (1995) found that *Allosyncarpia ternata* trees growing in the ravine had larger leaves with higher specific leaf area and had higher concentrations of chlorophyll *a* and *b* and had lower A_{max}, LCP, and quantum

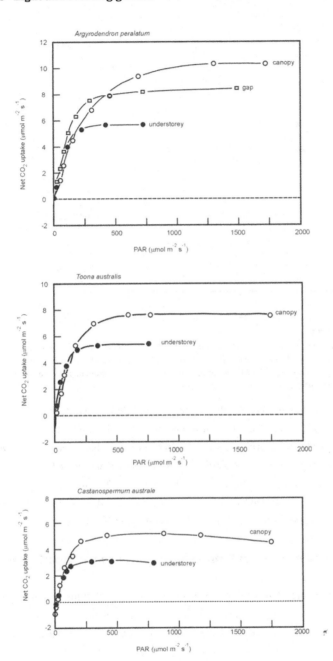

Figure 7.10

Relationship between photon flux density and assimilation of CO_2 by leaves of three Australian humid tropical rainforest tree species, Argyrodendron peralatum, Castanospermum australe *and* Toona australis. *For each species, comparisons are made between leaves growing in the understorey (solid circles), canopy (open circles) and, in the case of* Argyrodendron peralatum, *a forest canopy gap (open squares). Adapted from Pearcy (1987).*

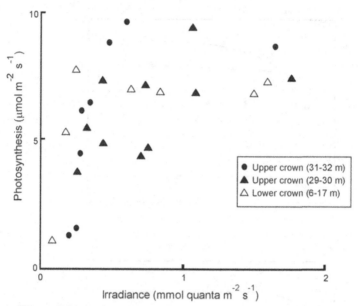

Figure 7.11

Relationship between light-saturated assimilation rate (A$_{max}$) and PAR for leaves on upper crown (circles), 2 m below crown (solid triangles) and lower canopy (15–26 m below crown; open triangles) on a 32 m tall Argyrodendron peralatum *tree. Measurements were taken one day after a rainfall event. Adapted from Doley* et al. *(1988).*

yield than *A. ternata* trees on the hill-top site (Table 7.3). Thus, in the cases of *Allosyncarpia ternata* and *Argryodendron peralatum*, there is clear evidence of physiologically and morphologically distinct 'sun' and 'shade' leaves. However, so great is the photosynthetic plasticity of some species that the dichotomization of 'sun' and 'shade' leaves breaks down. Doley *et al.* (1988) found no significant variation in A$_{max}$ between leaves from three positions in the canopy of the north Queensland rainforest tree *Argyrodendron peralatum* (Figure 7.11).

Maximum and minimum temperatures across the rainforest boundary

Minimum winter temperatures may set the altitudinal limit of subtropical rainforest and *Nothofagus moorei* rainforest (Fraser and Vickery 1938; Dodson and Myers 1986). Read and Hill (1988) also suggested that the distributions of some Tasmanian rainforest tree species are determined by frost tolerance. For example, *Atherosperma moschatum* is frost sensitive and thus restricted to low elevations while *N. cunninghamii* and *A. selaginoides* are more frost tolerant and can occur at high altitudes. Read and Hill (1989) experimentally determined the general order of frost resistance in a range of Australian temperate and subtropi-

Figure 7.12
*Deciduous beech (*Nothofagus gunnii*) covered by snow. The Tasmanian endemic N. gunnii,*
Australia's only winter-deciduous tree, occurs at the latitudinal and altitudinal limit of
Australian rainforest. (Photograph: Geoff Hope.)

cal rainforest trees. They suggested that the relative order of frost tolerance is as
follows:

Ceratopetalum apetalum = Doryphora sassafras ⩽ *Atherosperma moschatum* <
Nothofagus moorei = Eucryphia moorei < *Nothofagus cunninghamii* =
Eucryphia lucida < *Eucryphia milliganii.*

Sakai *et al.* (1981) found that the subtropical rainforest broad-leaved species
Nothofagus moorei and the subtropical–tropical conifer *Araucaria cunninghamii*
were less resistant to freezing temperatures than temperate rainforest species.
Nonetheless both these species tolerated temperatures below 0°C.

Comparative ecological data suggest that rainforest species have a greater
thermal range than most non-rainforest species (Figure 7.12). Howard and
Ashton (1973) noted that in Victoria, *Nothofagus cunninghamii* has an altitudinal range from sea level to the subalpine zone, implying wide thermal tolerance,
while five co-occurring *Eucalyptus* species exhibit clear altitudinal zonation
implying narrow thermal tolerance. The experimental evidence also confirms
that rainforest tree species are more tolerant of freezing temperatures than
non-rainforest tree species. Sakai *et al.*'s (1981) study demonstrated that temperate rainforest tree species, and particularly three Tasmanian conifers, were more
frost resistant than temperate non-rainforest species (Table 7.4).

TABLE 7.4. Freezing resistance temperatures (°C) of a range of
Australian rainforest and non-rainforest species

| Species | Temperature (°C) that causes tissue damage | | | | Rainforest species |
	Leaf	Bud	Cortex	Xylem	
Podocarpus lawrencei	−22	N.A.	−22	−22	X
Athrotaxis cupressoides	−20	−20	−20	−20	X
Diselma archeri	−17	−17	−20	−20	X
Microcachrys tetragona	−17	−17	−20	−20	X
Nothofagus gunnii	Deciduous	−17	−17	−17	X
Athrotaxis selaginoides	−17	−15	−15	−15	X
Eucalyptus pauciflora	−16	−20	−20	−23	
Eucalyptus gunnii	−15	N.A.	−18	−18	
Eucalyptus vernicosa	−13	−15	−15	−15	
Drimys tasmanica	−13	−15	−17	−13	X
Grevillea diminuta	−13	N.A.	−15	−17	
Araucaria bidwilli	−10	N.A.	−10	−10	X
Callitris oblonga	−10	−15	−15	−15	
Hakea lissosperma	−10	−10	−13	−15	
Leptospermum lanigerum	−10	−10	−10	−15	
Leptospermum flavescens	−8	−8	−10	−13	
Banksia marginata	−10	−10	−10	−10	
Callistemon rigidus	−8	−10	−10	−13	
Eucalyptus cinerea	−8	N.A.	−12	−12	
Callitris endlicheri	−8	−8	−10	−10	
Nothofagus moorei	−8	−8	−8	−12	X
Araucaria cunninghamii	−5	−5	−5	−5	X

N.A.: no available data.
Adapted from Sakai *et al.* (1981).

At the local level, frost may control the location and dynamics of the rainforest
boundary. In Tasmania, for example, frost may inhibit establishment of rainforest seedlings on exposed sites such as rainforest–grassland boundaries, or on sites
where rainforest is regenerating following a severe fire (Ellis 1964; Read and Hill
1988). It is thought that radiation frosts and trapping of cold air in topographic
depressions are important factors in controlling some subtropical and tropical
rainforest boundaries (Tommerup 1934; Fraser and Vickery 1938; Webb and
Tracey 1981). Duff and Stocker (1989) found that nine days of recurrent light

Figure 7.13
An abrupt Araucaria
bidwilli *boundary on the
Bunya Mountains in
southern Queensland.
Nocturnal surface
temperatures in the
grasslands can be up to 9°C
cooler than on the floor of
the adjacent forest.
(Photograph: David
Bowman.)*

frosts (> − 1.5 °C) damaged grass and pioneer species on a rainforest margin, although only some *Alphitonia petrei* and *Terminalia sericocarpa* individuals where killed. They suggested that rainforest boundaries are controlled by an interactive effect of frost with fire because frost-damaged vegetation is more prone to fire. Working on the Bunya Mountains in southeastern Queensland, Fensham and Fairfax (1996) found that some boundaries between *Araucaria bidwilli* and grasslands had steep minimum nocturnal temperature gradients (i.e. 0 °C versus − 9 °C respectively) associated with cold air ponding (Figure 7.13). However, they dismissed Webb's (1964) suggestion of the primacy of minimum temperatures in controlling the distribution of *Araucaria bidwilli*-dominated rainforest because many boundaries occur on ridges and steep slopes where cold air cannot accumulate. Their conclusion is also consistent with Sakai *et al.*'s (1981) finding that *Araucaria bidwilli* foliage is resistant to temperatures as low as − 10 °C (Table 7.4). In contrast to rainforest boundaries, there is solid evidence to support the view that frost limits the local distribution of *Eucalyptus* species (Harwood 1980; Davidson and Reid 1985) in the montane regions of southeastern Australia. The classic example of this are the 'inverted tree-lines' where *Eucalyptus* occurs up-slope from grasslands situated in drainage lines.

In the monsoon tropics, the rainforest boundary is characterised by differences in the maximum rather than minimum ground surface temperatures (Bowman and Panton 1993*b*). While there was no difference in mean minimum ground surface temperature (18.8°C versus 19.3°C), mean maximum ground surface temperatures were 4°C lower in the monsoon rainforest than the adjoining *Eucalyptus* savanna (38.9°C versus 43.2°C). Similarly, Bowman (1994) found that during the dry season, ground mid-day surface temperatures on either side of *Allosyncarpia ternata* rainforest boundaries differed by more than 10°C (30°C to over 40°C). It is possible that the low survival of *Allosyncarpia ternata* germinants in savanna is linked to moisture stress associated with high temperatures (Bowman 1991*a*; Bowman 1994). High temperatures and associated dryness is thought to adversely influence seed germination in large canopy gaps in humid tropical forests (Hopkins and Graham 1984*b*). However, it is unlikely that high ground surface temperatures affected the seedling survival of monsoon rainforest species. Transplanted seedlings of some monsoon rainforest species are known to tolerate the hot savanna microclimate (Bowman 1993), as can foliage on adult rainforest trees. For instance, foliage on adult *Allosyncarpia ternata* tolerated air temperatures in excess of 40°C (Fordyce *et al.* 1995) (Figure 7.14).

Conclusion

In this chapter, I have demonstrated that many Australian rainforest trees are tolerant of wide variations in light and temperature regimes. There is good evidence that minimum temperatures are important in controlling the distribution of different types of rainforest tree species across altitudinal and latitudinal gradients. In contrast, there is far less evidence that minimum temperatures are critical in controlling rainforest boundaries. A large body of research shows that many rainforest species are physiologically capable of growth under the high light and high temperature conditions characteristic of non-rainforest vegetation. This is consistent with the expansion of rainforest into adjacent *Eucalyptus* forests that has been observed in localities ranging from the humid tropics to Tasmania. Some non-rainforest plant species such as *Casuarina littoralis* are tolerant of low light conditions and have maximum growth in shade. Dense understoreys disadvantage light-demanding species such as *Eucalyptus*. The

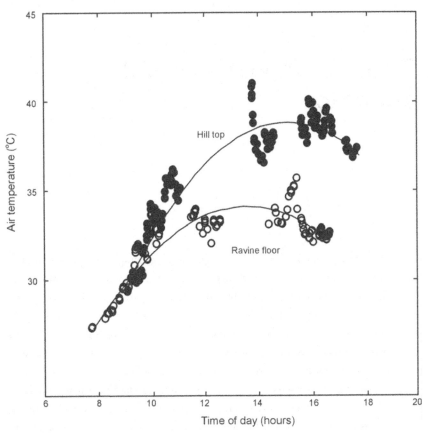

Figure 7.14
Comparison of the diurnal variation in air temperature in an Allosyncarpia ternata *forest growing in a moist ravine and a dry hill-top at the end of the dry season (November) in Kakadu National Park. Adapted from Fordyce et al. (1995).*

failure of *Eucalyptus* species to regenerate in large canopy gaps in both rainforest and non-rainforest environments does not appear to be solely related to light regime, given that rainforest species that are as light demanding as *Eucalyptus* can successfully establish in these shaded environments. Other factors such as pathogens, seedbed characteristics, competition, and seed dispersal appear to be important in preventing *Eucalyptus* from regenerating in rainforest canopy gaps (Ashton and Willis 1982).

Fire is known to create conditions that favour the regeneration of *Eucalyptus,* and conversely the development of dense understoreys beneath *Eucalyptus* canopies can only occur in the absence of fire disturbance. Therefore the steep microclimatic gradient across the rainforest boundary must be considered as a consequence rather than the cause of the rainforest boundary.

8

The fire theory I. Field evidence

Numerous researchers working throughout the geographic range of Australian rainforests have advanced the theory that rainforest boundaries are controlled by fire. This is an old idea, possibly first advanced by the Czech botanist Domin (1911). At the heart of the theory is the assumption that rainforest tree species are 'fire tender'. Francis (1951) wrote that 'one of the most marked differences between the constituents of rainforests and those of the open *Eucalyptus* forests is their behaviour towards fire. In most if not all cases the rainforest constituents are killed even by slight contact with or proximity to the fires which periodically sweep through many of the *Eucalyptus* and open forests of Australia'. Similarly, Webb (1968) argued that the difference in the ability of rainforest and non-rainforest vegetation to tolerate fire 'is reflected in the remarkably sharp boundaries of fire-sensitive raingreen forests in the tropics and subtropics, which is related to the exclusion of fire, virtually on an all-or-nothing basis'. However, the supporting evidence is limited and often circumstantial. My aim in this chapter is to review the field evidence that fire is critical in controlling rainforest boundaries throughout Australia. Evidence from the humid tropics, monsoon tropics, subtropics and temperate regions will be considered in turn.

Humid tropics

Webb (1968) argued that, in the tropics, rainforest and non-rainforest vegetation is in a dynamic balance controlled by fire frequency, soil fertility and topographic settings such as 'rocky outcrops and gullies, especially in the lee of fire-bearing winds'. Building on this work, Ash (1988) developed a generalised model to explain the role of rainfall, soil fertility and topography in determining the stability of rainforest boundaries in the Australian humid tropics (Figure 8.1).

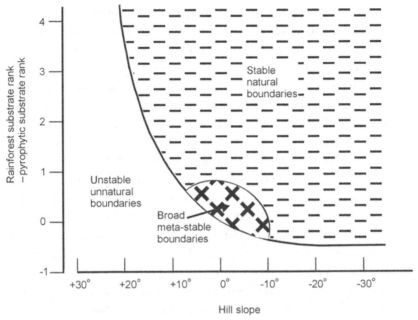

Figure 8.1
General model to explain the stability of humid tropical rainforest boundaries on the Atherton Tablelands in north Queensland. The x-axis shows the slope angle of rainforest boundaries. Negative values indicate rainforest down-slope from non-rainforest vegetation and positive values indicate rainforest up-slope from non-rainforest vegetation. The y-axis is a relative measure of the magnitude of the difference in soil fertility. It is determined by subtracting the ordinal scores of fertility for different geologies between rainforest and non-rainforest vegetation where: stony basalt: 5; basalt: 4; alluvium: 3; metamorphics: 2; granite: 1. The solid line indicates the natural occurrence of rainforest boundaries. Narrow (20 m wide), and presumably stable, boundaries occur in most of the area above the solid line. Broad (20 to 200 m wide) rainforest boundaries occur in the hatched area. Apparently unstable boundaries occur outside the solid line in unnatural landscape settings such as above cane fields. Adapted from Ash (1988).

The model was based on data compiled from maps of vegetation, rainfall, geology, and topography in order to identify trends and relationships between these variables across rainforest boundaries. Although the data set was unsuitable for statistical analysis due to numerous missing values, Ash was able to identify 'tentative rules for determining where a boundary is most likely to occur with a particular combination of factors, and whether such a boundary is likely to be stable'. Neither geology nor rainfall was found to explain the location of all rainforest boundaries in his study region, although Ash found an interaction between these variables such that rainforest in lower rainfall areas typically occurred on fertile soils derived from basalt (Figure 6.6). In most circumstances, the rainforest was down-slope of non-rainforest vegetation. He attributed this distribution not only to the accumulation of moisture, nutrients and soil down-slope, but also to the geometric effect of slope on fire intensity. When fires burn downhill they are of lower intensity than when they burn uphill (Ash 1988; Unwin *et al.* 1985).

Nonetheless, Ash (1988) noted the occurrence of some rainforests up-slope from non-rainforest vegetation. Typically this occurred at high elevation sites with gentle slopes and deep fertile soils and which were frequently enveloped in cloud. He noted that rainforests were generally absent from ridges where exposure to winds causes desiccation and there is a higher probability of fire caused by lightning strikes. Where rainforest occurs on a more fertile substrate than the adjacent non-rainforest, Ash (1988) argued that the greater the difference between substrate fertility the greater the likelihood that the boundary will be stable. However, he suggested that slope angle also influenced this relationship. Once slope angle exceeds 20° the risk of fire is thought to be so great that it overrides any effect of soil fertility in favouring the development of rainforest. Ash acknowledged that rocky or bare ground can act as a barrier to fire, enabling rainforest to occur on steep slopes and infertile soils. Also contrary to the prediction of his model is the observation of Unwin (1989) that the position of rainforest boundaries was not consistently related to slope angle even on sites with uniform geology. The cause for this imperfect relationship may be related to the broader topographic setting of a site. Local features such as hills and creek lines can act as topographic barriers to winds, and thus protect rainforest from fire. Alternatively they could intensify windstorms (Webb 1958), channel or pond freezing air and expose slopes to 'fire-bearing winds' (Webb 1968; Unwin *et al.* 1985).

Ash's model is based on the fundamental assumption that rainforest species are incapable of surviving fire. Indeed he described rainforest plants as being 'pyrophobic'. This view needs qualification, however. Experiments have shown that many humid tropical rainforest trees can recover from at least a single fire. Unwin *et al.* (1985) found that canopy scorch resulting from fires of low to moderate intensity declined across a rainforest boundary, and that half the fire-damaged trees recovered by coppicing (Table 8.1). These findings are consistent with those of Stocker (1981), who found that the dominant mode of regeneration of trees in a cleared and burnt humid tropical rainforest was by coppicing from stumps (Table 8.2). The limited role of seed regeneration reported by Stocker (1981) was probably due to the death of seeds on exposure to high soil temperatures (Hopkins and Graham 1984a). They found that compared with controls, the germination rate and species diversity of humid tropical

TABLE 8.1. The effect of low and moderate intensity fires on crown scorch and mortality of small scorched trees and shrubs across a rainforest boundary on the Atherton Tablelands

Scorch was defined as the death of more than half the foliage in the crown. Fire intensities were: low: 240–370 kW m^{-1}; moderate: 780–2350 kW m^{-1}

	% scorched (no. of plants)		% mortality (no. of stems)	
Vegetation type	Low	Moderate	Low	Moderate
Eucalyptus forest/ ecotone edge	92 (26)	92 (63)	26 (24)	29 (58)
Ecotone	50 (44)	68 (38)	29 (22)	42 (26)
Ecotone/rainforest edge	18 (84)	33 (45)	47 (15)	27 (15)

Adapted from Unwin *et al.* (1985).

TABLE 8.2. Regeneration modes of 82 tree species following logging and burning of a humid tropical rainforest on the Atherton Tablelands

Regeneration mode	Percentage (number) of species
Seeds only	9.9 (8)
Suckers only	2.5 (2)
Coppicing only	52.5 (43)
Seeds and suckers	0 (0)
Seeds and coppicing	25.6 (21)
Suckers and coppicing	3.7 (3)
Seeds, suckers and coppicing	6.1 (5)
Total seeds	41.6 (34)
Total suckers	12.3 (10)
Total coppicing	87.9 (72)

Adapted from Stocker (1981).

rainforest soil seed banks was dramatically reduced when heated at 60 °C for one hour. Heating to 100 °C for one hour resulted in almost complete germination failure.

The capacity of most humid tropical rainforest species to recover vegetatively from fire damage suggests that low-intensity fires would not cause the substantial retreat of rainforest boundaries unless they occurred frequently enough to exhaust reserves of carbohydrate so that trees can no longer re-sprout. However, high intensity fires following extreme weather events (droughts, frosts or severe storms) which produce large masses of fuel are thought to be capable of destroying tropical rainforests (Webb 1958; Ridley and Gardner 1961; Unwin *et al.* 1985). For instance, there is a high risk of intense fires after a tropical storm because destruction of the forest canopy leads to heavy fuel loads, high temperatures and low humidity on the forest floor (Webb 1958; Stocker 1981; Hopkins and Graham 1987; Unwin *et al.* 1988; Turton 1992). The dense crowns of rainforest trees appear to be more susceptible to storm damage than adjacent open crowned non-rainforest vegetation (Unwin *et al.* 1985, 1988). Unwin *et al.* (1988) suggested that fires in storm debris are a 'probable means of conversion to grassy open forest, especially where the effects of cyclone and subsequent drought are combined'. Fires in storm debris may permit the establishment of a grass layer, increasing the risk of further fires that may ultimately exhaust the capacity of rainforest trees to recover. The result is conversion of rainforest to non-rainforest vegetation (Stocker and Mott 1981).

The interaction of frosts and fire is also thought to control the distribution of rainforest. On exposed rainforest margins, frosts kill vegetation and increase the available fuel. Duff and Stocker (1989) suggested that fires driven by the prevailing southeasterly winds in fuels created by frosts associated with the ponding of cold air in topographic depressions may account for islands of *Eucalyptus* forest on the western and northwestern margins of swampy sites in areas otherwise dominated by rainforest (Figure 8.2).

There is clear evidence that the absence of fire on rainforest margins permits their expansion. Unwin (1989) monitored the demographic structure of trees in plots across a rainforest boundary on the Atherton Tablelands over a 10-year period. During this time, he observed the expansion of rainforest species such as *Flindersia brayleyana*, *Neolitsea dealbata* and *Acacia aulacocarpa* beneath the *Eucalyptus grandis* forests that form an ecotone on the rainforest margin. There was a concurrent decline in grasses such as *Imperata cylindrica* and *Themeda australis*, and the bracken fern *Pteridium esculentum* in the ground layer, these being replaced by the more shade-tolerant robust sedge *Gahnia aspera*, the ginger *Alpinia coerulea* and numerous tree seedlings, shrubs and vines. Unwin estimated the advance of the rainforest to be about 1 m per year. These findings account for the widespread occurrence of *Eucalyptus* trees emergent over an understorey of young tropical rainforest trees (Ridley and Gardner 1961; Unwin 1989; Harrington and Sanderson 1994). Harrington and Sanderson's analysis of

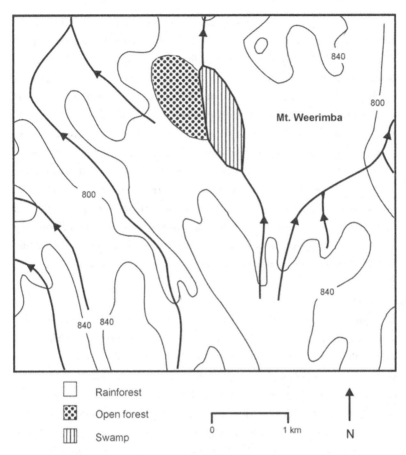

Figure 8.2
Vegetation map showing an isolated patch of Eucalyptus *within a large tract of rainforest on the Atherton Tablelands. The* Eucalyptus *forest is to the northwest of a swamp formed when the Creek was blocked by a lava flow. The major drainage lines and 40-m contours are shown. Adapted from Duff and Stocker (1989).*

aerial photographs taken between 1943 and 1992 showed that rainforest had invaded 70% of the tall *Eucalyptus grandis* forests and 57% of mixed-species tall *Eucalyptus* forest in three study areas in northeast Queensland. The recent expansion of rainforest into adjoining *Eucalyptus* forests is thought to be related to the reduced frequency and intensity of grass fires associated with livestock grazing and the consequent reduction in grassy fuels, and possibly also to the concurrent cessation of Aboriginal landscape burning (Unwin *et al.* 1985; Unwin 1989; Harrington and Sanderson 1994). Conversely, Unwin *et al.* believed that high grass biomass and associated frequent fires prevented the expansion of rainforest into otherwise favourable sites.

In summary the boundary of tropical rainforests is determined by the interaction of topography, soil fertility and fire history. Given that humid tropical rainforests are relatively resistant to mild fires, rainforest is most likely to retreat if fires occur with high frequency due to a high biomass of grass. However, a single fire can cause the rapid retreat of rainforest when extreme climatic events such as frosts, droughts and cyclones produce large accumulations of dry fuels. In the absence of fire, rainforest seedlings can invade adjacent non-rainforest vegetation. Once the invading rainforest has achieved canopy closure, the consequent elimination of grassy fuels decreases the risk of fire, which in turn favours the long-term establishment of rainforest. The critical factor in controlling the distribution of rainforest in the Australian humid tropics is the periodicity and intensity of fire (Figure 8.3). Given the infrequent occurrence of lightning storms, humans appear to be an important source of ignitions, both currently and in pre-European times (Ash 1988).

Monsoon tropics

A perplexing feature of the Australian monsoon tropics is the small patch size and widely scattered distribution of rainforest. Climatic analyses led Mackey *et al.* (1989) to conclude that deciduous vine thicket rainforest occupies a tiny proportion of its potential geographic range. Similarly, at the conclusion of an extensive study of monsoon rainforest throughout the Northern Territory, Russell-Smith (1991) wrote: 'given the range of moisture conditions and landforms which monsoon rainforest vegetation currently occupies in northern Australia, the fundamental question remains: why is the distribution of rainforests so restricted in that region?'

A recurrent explanation has been the role of fire. Some authors believe fire has caused the breakdown of a formerly extensive rainforest, while the alternative view is that the rainforests have opportunistically colonised fire-protected sites. There is evidence for both views, and these perspectives are not mutually exclusive (Russell-Smith 1985a). The issue is complex, however, because types of monsoon rainforest differ in their response to fire and protection from it.

The best evidence of the breakdown of a formerly extensive tract of monsoon rainforest by fire concerns the forests dominated by *Allosyncarpia ternata* which constitute over 40% of the rainforest in northwestern Australia. This rainforest type is thought to have a relictual distribution because the genus is monotypic and endemic to the rugged western edge of the Arnhem Land Plateau, and has a heavily fragmented distribution (Figure 8.4) (Bowman 1991a; Russell-Smith *et al.* 1993). The nearest relatives of *A. ternata* are distinct genera from New

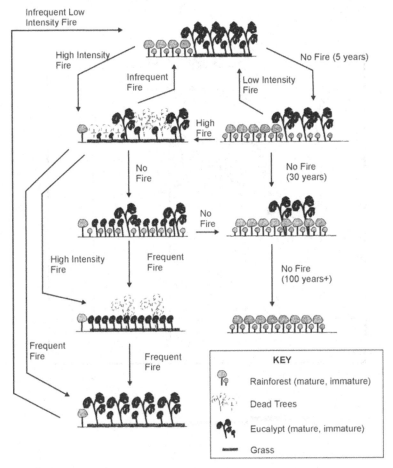

Figure 8.3
Flow diagram showing humid tropical rainforest boundary dynamics as controlled by fire frequency and intensity. Adapted from Harrington (1995).

Caledonia (*Arillastrum*), eastern Malesia (*Eucalyptopsis*) and eastern Australia (an undescribed taxon with a very restricted distribution). *Arillastrum*-type taxa including *Allosyncarpia* are thought to be primitive groups that have subsequently diversified into *Eucalyptus* (Ladiges *et al.* 1995). Although there is no fossil record of *A. ternata*, biogeographic evidence points to it being a Tertiary relict like *Nothofagus cunninghamii*, for which there is a good fossil record (Hill 1983). The fragmented regional and local distribution of *A. ternata* is curious given the

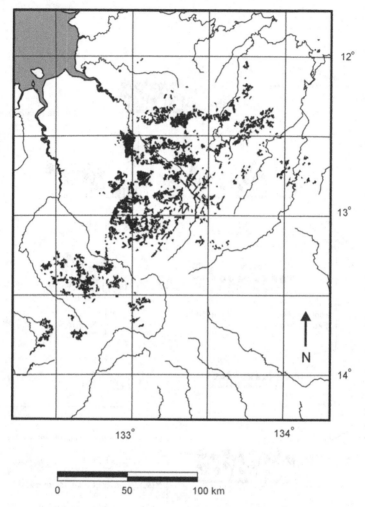

Figure 8.4
The global distribution of Allosyncarpia ternata-*dominated monsoon rainforests that are endemic to the western edge of the Arnhem Land Plateau, Northern Territory. Adapted from Russell-Smith et al. (1993).*

species' wide ecological tolerance (Bowman *et al.* 1993; Russell-Smith *et al.* 1993; Fordyce *et al.* 1995, 1997*a*), and is linked to its sensitivity to fire. Although *A. ternata* can recover vegetatively, fire is a common cause of mortality particularly of large trees (Bowman 1991*b*, 1994). The capacity of *A. ternata* to re-colonise areas where fire has eliminated the species is limited because of poor seed dispersal (Bowman 1991*a*). It is possible that burning throughout the Quaternary, or indeed the late Tertiary, caused the fragmentary distribution of the

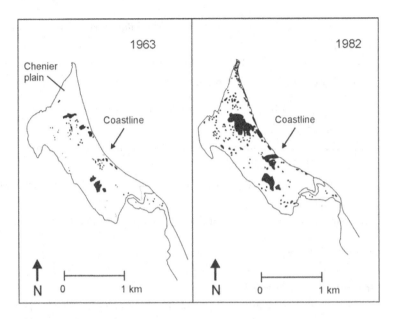

Figure 8.5
The colonisation of treeless chenier plains by monsoon rainforest on Cobourg Peninsula, Northern Territory, between 1963 and 1982. Adapted from Bowman et al. *(1990).*

species. However, there is evidence, to be discussed in Chapter 11, that Aboriginal use of landscape fires conserved *A. ternata* rainforests (Russell-Smith *et al.* 1993).

In contrast to the *A. ternata* rainforests, the fragmented distribution of coastal monsoon rainforests is attributed to their capacity to colonise recently formed landscapes (Fensham 1993). Fifteen per cent of 1219 patches of monsoon forest sampled by Russell-Smith and Lee (1992) in the Northern Territory were on coastal dunes and floodplain margins that had formed in the Holocene. Bowman *et al.* (1990) demonstrated that over a 20 year period, rainforest had established on grassy chenier plains on Cobourg Peninsula in the Northern Territory (Figure 8.5). Rainforest clumps on these coastal plains were dominated by pioneer tree species such as *Pandanus spiralis, Acacia auriculiformis, Alstonia actinophylla, Timonius timon* and *Casuarina equisetifolia*, which provide shelter for the establishment of a wider variety of monsoon rainforest tree species. About half of the 32 woody species recorded in the clumps were only represented by juveniles. Fire can destroy the clumps, and repeated fires in the past are thought to account for the existence of stunted and isolated rainforest trees such as *Alstonia actinophylla, Timonius timon* and *Ficus opposita*, species that are relatively fire resistant owing to their capacity to re-sprout from root stocks. Bowman *et al.* concluded that the colonisation by, and destruction of, monsoon rainforest patches is controlled by fire frequency.

Over the last two decades, a huge research effort has greatly expanded our knowledge of monsoon rainforests, leading to the following general conclusions:

(i)   there is a correlation between monsoon rainforest distribution and topographic fire protection;

(ii) the occurrence of monsoon rainforests without topographic fire protection is unusual;

(iii) monsoon rainforest trees have a capacity to recover following damage by fire; and

(iv) monsoon rainforest cannot colonise long unburnt *Eucalyptus* savanna.

Each will be discussed in turn.

Topographic fire protection

Surveys of rainforest in the Northern Territory and the inland regions in northeastern Queensland show a strong positive relationship between rainforest occurrence and sites with topographic protection (Russell-Smith 1991; Fensham 1995) (Figure 8.6). However, both workers noted exceptions to this rule. In a study of 144 rainforest savanna boundaries throughout northwestern Australia, Bowman (1992a) found that monsoon rainforest sites were differentiated from surrounding savannas in terms of site rockiness and slope angle (Table 8.3). The magnitude and direction of these differences was found to depend upon the type of monsoon rainforest. Wet monsoon rainforests (i.e. on springs and other perennial soil moisture supplies) occurred on more level terrain and were less rocky than surrounding savannas. The permanently moist conditions in wet monsoon rainforest were thought to inhibit wildfire. In contrast, dry monsoon rainforests occurred on steeper and rockier terrain than surrounding savannas. Bowman concluded that dry rainforests were typically restricted to steep rocky sites because rocks are often effective firebreaks. Fordyce *et al.* (1997b) documented a decrease in fire intensity up a rocky slope (Figure 8.7), this being an exception to the rule that fires are more intense when they burn uphill (Ash 1988). Although rocky dry hills provide rainforest with some fire protection (Kirkpatrick *et al.* 1988), sheltered gorges are clearly the most fireproof environments. In addition to fire protection, gorges often provide enhanced moisture supplies and are therefore particularly favourable to rainforest. This correlation has often been erroneously interpreted as demonstrating a causal link between rainforest and enhanced supplies of soil moisture.

Figure 8.6
A deep ravine provides topographic fire protection for a rainforest in the Australian monsoon tropics. (© Murray Fagg, Australian National Botanical Gardens.)

TABLE 8.3. Rock cover and slope (mean, coefficient of variation) in quadrats placed on 144 wet and dry monsoon rainforest boundaries in northwestern Australia

Vegetation type	Rock cover (%)	Slope angle (°)
Wet monsoon rainforest	7.7 (282)	4.0 (170)
Wet monsoon rainforest ecotone	22.9 (141)	10.1 (112)
Wet savanna	17.8 (152)	8.6 (128)
Dry monsoon rainforest	31.6 (115)	9.1 (111)
Dry monsoon rainforest ecotone	30.5 (118)	9.0 (113)
Dry savanna	22.0 (140)	6.6 (138)

Adapted from Bowman (1992a).

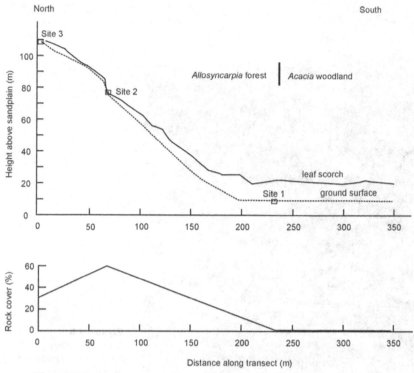

Figure 8.7
The height of leaf scorch following a low intensity (e.g. 500–3000 kW m⁻¹) late dry season fire that burnt across an Allosyncarpia ternata *boundary in Kakadu National Park on 1 August 1993. The percentage cover of the ground surface with rocks is also shown. Adapted from Fordyce* et al. *(1997b).*

Level terrain

Monsoon rainforests can occur without any topographic fire protection, for example on basaltic or lateritic plateaux or sand sheets (Langkamp *et al.* 1981; Bowman 1991*a*, 1992*a*; Bowman *et al.* 1990; Fensham 1995) (Fig. 1.17), phenomena for which there are a number of possible explanations. Fensham (1995) noted that 'dry rainforests' without topographic fire protection typically occur on fertile and often basaltic soils. Fensham accepted Webb and Tracey's (1981) conjecture that fertile soils favour grass understoreys that are frequently burnt by low-intensity fires, and that these conditions contrast with the infrequent, high-intensity fires associated with shrubby understoreys on less fertile soils. Low-intensity grass fires are thought to have little effect on the boundaries of rainforests growing on level terrain. There is evidence that differences in fuel characteristics across monsoon rainforest boundaries are a barrier to low inten-

TABLE 8.4. Biomass of grass and leaf litter, and moisture content early
(May) and late (September) in the dry season across a dry monsoon
rainforest boundary

Values are mean (range) and notation of values not significantly different at $P = 0.05$
(Mann–Whitney U-tests)

Component	Eucalyptus savanna	Monsoon rainforest ecotone	Monsoon rainforest
Grass (g m^{-2})	220 (85–412)	4 (0–18)	0 (0)
Leaf litter (g m^{-2})	269 (91–562) a	541 (202–728) b	452 (247–692) b
May moisture content of litter layer (%)	5.4 (3–9)	12.1 (8–17)	18.6 (14–24)
September moisture content of litter layer (%)	2.6 (1–4)	6.1 (4–8)	9.9 (6–19)

Adapted from Bowman and Wilson (1988).

sity fires. Bowman and Wilson (1988) found that compared to adjacent *Eucalyptus* savanna, monsoon forests had less grass, more leaf litter and higher fuel moisture content. However, the effectiveness of this fire barrier diminished with the time elapsed since the summer rains as the fuels were desiccated, rendering rainforests on level terrain vulnerable to damage by fire in the late dry season (Gill *et al.* 1996) (Table 8.4). Should a fire penetrate a rainforest, the risk of subsequent fires is increased because burning triggers the germination of seeds of herbaceous plants held in the soil seed bank (Russell-Smith and Lucas 1994). Bowman (1994) documented the contraction of an *Allosyncarpia ternata* rainforest growing on a sand sheet because of a high frequency of intense fires late in the dry season.

Recovery following fire damage

Burnt tree trunks and charred logs are conspicuous proof that both dry and wet monsoon rainforest can recover from fire (Bowman and Dunlop 1986; Bowman and Minchin 1987) (Table 8.5). Most rainforest trees can recover vegetatively from damage inflicted by a single event. Bowman (1991*b*) found that a late dry season fire resulted in the death of only 11% of 1271 stems sampled from both wet and dry monsoon rainforests, and that there were no deaths in 21 of the 32 species in the sample. Smaller stems were more likely to be damaged or killed by fire, with the exception of *Allosyncarpia ternata*. These results have been corroborated by

TABLE 8.5. The density of charred logs and tree trunks across a dry
monsoon rainforest boundary

Vegetation type	Charred logs per ha	Charred tree trunks per ha
Eucalyptus savanna	200	240
Monsoon rainforest ecotone	306	69
Monsoon rainforest	100	27

Adapted from Bowman and Dunlop (1986).

other studies of tagged individuals, as well as experiments that determined the
response of monsoon rainforest seedlings to different fire intensities. Only 22% of
mature *Ptychosperma bleeseri*, a rare palm restricted to wet monsoon rainforests,
were killed by a wildfire, and many fire-damaged individuals re-sprouted from
root stocks (Barrow *et al.* 1993). Experimental fires of various intensities demonstrated that seedlings of most monsoon rainforest tree species tested recovered
vegetatively (Table 8.6), exceptions being *Allosyncarpia ternata* and *Callitris
intratropica* (Bowman 1991*b*; Bowman and Panton 1993*a*). Fordyce *et al.* (1997*b*)
found, however, that the survival of burnt *Allosyncarpia ternata* seedlings increased with age. All seedlings less than 1.5 years old were killed by a low to
medium intensity fire (e.g. $0.5–3.0$ MW m^{-1}), while 87% of seedlings more than
3.5 years old survived the fire. Similar trends were evident amongst burnt 0.5 to 5
year old *Allosyncarpia ternata* seedlings raised in a shade house (Fordyce *et al.*
1997*b*). Russell-Smith (1996) found that about a third of the 131 monsoon
rainforest tree species and half the 48 shrub and 57 vine species he sampled
produced clones, which may contribute to the capacity of many monsoon
rainforest plants to recover vegetatively from fire.

 Given the capacity of many monsoon rainforest trees to recovery following
burning, repeated fires are required to cause boundary retreat (Stocker and Mott
1981; Bowman 1991*b*). An exception to this is the combined impact of severe
tropical storms and subsequent fire in the debris. Bowman and Panton (1994)
found that a fire that burnt through a wet monsoon rainforest killed a higher
proportion of small stems, while in a dry monsoon rainforest a cyclone caused
damage to large stems. They suggested that a combination of a cyclone followed
by a fire in the debris could have a significant impact on a rainforest because of
the destruction of all tree size classes (Table 8.7). Rapid contraction of monsoon
rainforest following severe tropical storms has been documented (Panton 1993;
Bowman *et al.* 1999*a*) (Figure 8.8).

TABLE 8.6. Recovery of seedlings less than 1 year of age of 14 monsoon rainforest tree species 2 months after fires in a range of grass fuel loads (percentage, with no. of individuals in treatment in parentheses). The fires were lit at midday in September 1990 under calm, cloudless conditions

| Species | Grass fuel mass (mg ha^{-1}) | | | |
	2	4	6	8
Acacia auriculiformis	30 (10)	0 (10)	0 (11)	0 (11)
Allosyncarpia ternata	0 (7)	0 (8)	0 (6)	0 (9)
Aidia cochinchinensis	20 (10)	25 (12)	8 (12)	0 (11)
Bombax ceiba	50 (8)	50 (8)	38 (8)	25 (8)
Breynia cernua	71 (7)	67 (9)	71 (7)	44 (9)
Callitris intratropica	0 (5)	0 (12)	0 (6)	0 (7)
Drypetes lasiogyna	18 (11)	17 (6)	11 (9)	0 (8)
Helicteres isora	100 (11)	83 (12)	92 (12)	91 (11)
Micromelum minutum	0 (13)	13 (8)	0 (12)	0 (11)
Polyalthia nitidissima	0 (12)	0 (9)	10 (10)	11 (9)
Sterculia quadrifida	80 (5)	100 (6)	83 (6)	83 (6)
Strychnos lucida	100 (7)	50 (8)	14 (7)	86 (7)
Terminalia muelleri	20 (10)	0 (10)	0 (9)	0 (9)
Vitex acuminata	100 (12)	100 (12)	64 (11)	77 (13)

Results for *Bombax ceiba and Sterculia quadrifida* were reported by Bowman and Panton (1993*b*), for *Allosyncarpia ternata* by Bowman (1991*a*) and *Callitris intratropica* by Bowman and Panton (1993*a*). The results for the remaining 10 species are from my unpublished notes.

TABLE 8.7. Size of monsoon rainforest tree stems damaged by fire and cyclone

Sizes are diameters at breast height (DBH). Values are mean (range) and sample size

Disturbance	DBH of undamaged stems, in cm	DBH of damaged stems, in cm
Fire	7.6 (1–100) 3398	4.1 (1–45) 598
Cyclone	4.0 (1–43) 598	12.9 (1–56) 165

Adapted from Bowman and Panton (1994).

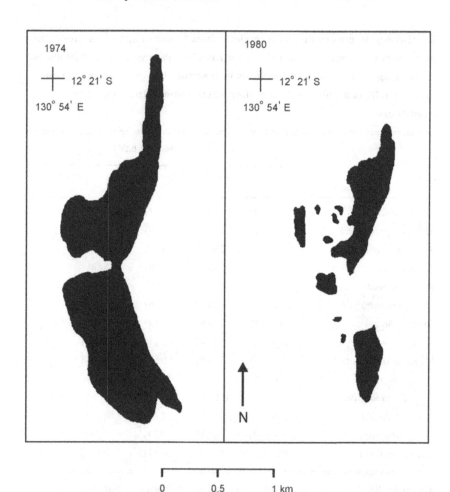

Figure 8.8
*Contraction of rainforest near Darwin following a severe tropical cyclone on Christmas Day
1974. Adapted from Panton (1993) and Bowman et al. (1999a).*

Unburnt savanna

The vast expanse of *Eucalyptus* savanna in monsoonal northern Australia has
been considered by some workers as secondary or disclimax forest produced by a
long history of Aboriginal burning and maintained by the current high frequency
of landscape fires (Stocker and Mott 1981; Kershaw 1985; Russell-Smith *et al.*
1997*b*). A corollary of this view is that relief from burning would allow rainforest
to colonise the savanna (Langkamp *et al.* 1981). However, studies demonstrate
the extremely limited capacity of rainforest to colonise unburnt savanna. For
example, Fensham (1990) found low densities (one seedling per 0.2 ha) of only

TABLE 8.8. Density of seedling savanna, ecotone and rainforest tree species across a rainforest–*Eucalyptus* savanna boundary protected from fire for more than 15 years at Weipa on Cape York Peninsula, Queensland

A seedling was defined as being less than 1.5 m tall. Values are mean (standard error) seedlings per 2 m²

Seedling type	Vegetation type (seedlings per 2 m²)		
	Eucalyptus savanna	Ecotone	Monsoon rainforest
Eucalyptus savanna	1.1 (0.19)	0.5 (0.17)	0
Ecotone	0.2 (0.07)	0.9 (0.35)	0
Monsoon rainforest	0.1 (0.05)	1.4 (0.44)	9.5 (1.6)

Adapted from Bowman and Fensham (1991).

three monsoon rainforest species (*Timonius timon, Denhamia obscura* and *Alstonia actinophylla*) that had established in a *Eucalyptus* savanna protected from fire for 10 years. Bowman and Panton (1995) documented the failure of any rainforest species to invade a *Eucalyptus* savanna that had been protected from fire for 20 years. It is important to put these studies into context. Under the current biennial burning regime the probability of a savanna escaping fire for 20 years is one in a million. The persistence of dry grass fuels in long-unburnt savannas keeps the risk of fire high, rendering rainforest succession almost impossible under the current fire regime (Bowman *et al.* 1988*b*; Bowman and Panton 1995).

It is possible that the scarcity of rainforest seedlings in unburnt savannas is attributable to an insufficiency of seed dispersal, though many monsoon rainforest tree seeds are dispersed by wind or mobile fauna such as birds and bats. However, this hypothesis is not supported by studies of two unburnt rainforest boundaries where seed dispersal is unlikely to be limiting. Bowman and Fensham (1991) found that, after protection from fire for 15 years, a narrow ecotone dominated by microphyllous shrubs such as *Bridelia tomentosa* and *Callicarpa pedunculata* had developed on the rainforest edge. Seedlings of monsoon forest trees had established at low densities in the unburnt savanna and at much higher densities beneath the ecotone (Table 8.8). They suggested that the establishment of rainforest seedlings appears to depend on environmental conditions created by the development of an ecotone. Similarly, Bowman (1993) found only very

TABLE 8.9. Frequency of *Aglaia rufa* and *Diospyros maritima* seedlings in 4 m² permanent quadrats in a monsoon rainforest ecotone and *Eucalyptus* savanna protected from fire between 1988 and 1992

Vegetation type	Aglaia rufa		Diospyros maritima		Number of quadrats
	1988	1992	1988	1992	
Eucalyptus savanna	0	0	0	3	24
Monsoon rainforest ecotone	2	1	4	7	26

Adapted from Bowman (1993).

low rates of establishment over four years of two locally abundant monsoon rainforest trees (*Aglaia rufa* and *Diospyros maritima*) in an unburnt rainforest ecotone and adjacent unburnt *Eucalyptus* savanna (Table 8.9). There is experimental evidence that suggests that some monsoon rainforest tree species may be prevented from establishing in *Eucalyptus* savanna because of unfavourable soil biota (Stocker 1969; Bowman and Panton 1993b). Sterilisation of monsoon rainforest soil limited stem growth of the rainforest tree *Bombax ceiba* to the same level as on sterilised and unsterilised *Eucalyptus* savanna soil, but the response of *Sterculia quadrifida* was less pronounced (Table 8.10).

Summary

Rainforest in the monsoon tropics is mostly restricted to landscape settings protected from fire. Higher levels of moisture in fuels in the early dry season protect rainforests from savanna fires, but this protection breaks down as the dry season progresses. Monsoon rainforest tree species have a well developed capacity to recover vegetatively following damage by fire. There is evidence to suggest that fire protection enables rapid expansion of monsoon rainforest on geologically young and treeless landscapes (i.e. margins of floodplains, creek lines, coastal plains, coastal dunes), but expansion into unburnt *Eucalyptus* savannas is slow. The cause of this differential response is unclear and may be related to competition with the *Eucalyptus* savanna and unsuitable soil biota (Bowman and Panton 1995). The extremely high frequency of fires in savannas and the slow rate of rainforest expansion into unburnt *Eucalyptus* savannas makes the concept of secondary rainforest succession unrealistic in the Australian monsoon tropics (Bowman 1988; Fensham 1990; Bowman and Panton 1993b).

TABLE 8.10. Germination and seedling stem growth of *Bombax ceiba* and *Sterculia quadrifida* after 120 days in monsoon rainforest and *Eucalyptus* savanna with or without methyl bromide sterilisation

Values are mean (standard error) and notation of means not significantly different at $P = 0.05$ (SNK multiple range tests). Percentage germination rates were compared using a square root $(x + 0.5)$ transformation and heights compared using a $\log_e (x + 1)$ transformation

Growth response	Species	Monsoon rainforest		*Eucalyptus* savanna	
		Unsterilised	Sterilised	Unsterilised	Sterilised
Germination (%)	*Sterculia quadrifida*	56.6 (8.8) a	15.0 (8.7) b	52.5 (6.3) a	15.0 (5) b
	Bombax ceiba	32.5 (10.3) c	2.5 (2.5) d	10.0 (4.1) c	2.5 (2.5) d
Stem growth (mm)	*Sterculia quadrifida*	198 (14) e	135 (15) ef	115 (3) f	123 (7) f
	Bombax ceiba	429 (26) g	120 (9) h	154 (12) h	110 (7) h

Adapted from Bowman and Panton (1993*b*).

Subtropics

Florence (1964,1965) argued that the distribution of rainforest and *Eucalyptus* forest in the subtropics is controlled by a complex of edaphic factors including soil fertility, soil permeability and soil moisture, and that fire is a superimposed influence on the primary importance of these edaphic variables. Florence (1965) wrote that 'were fire a major influence on vegetation pattern, consistent vegetation–environment patterns would not be readily demonstrated'. Nonetheless, a number of workers have observed that various rainforest tree species are capable of invading *Eucalyptus* forests in the subtropics (Herbert 1936; Blake 1938; Fraser and Vickery 1939; Cromer and Pryor 1943; Burges and Johnston 1953; Smith and Guyer 1983).

Although some authors believed the expansion of rainforest was linked to climatic change (Fraser and Vickery 1939; Burges and Johnston 1953), most accepted that fire controlled the position of the rainforest boundary. For example, Herbert (1936) recorded the colonisation of *Eucalyptus* forests on north-facing slopes by *Nothofagus moorei* in the highlands of southeastern Queensland and concluded that 'there seems no reason why they should not reach maturity, provided bush fires do not sweep the area before they form a closed community'. Similarly, Osborn and Robertson (1939) noted that in the Myall Lakes region in NSW, rainforest could expand if it was protected from fire. Cromer and Pryor (1943) mapped species distributions across the boundaries of an *Araucaria cunninghamii*-dominated rainforest at Widgee in southern Queensland. They found that some of the boundaries between rainforest and *Eucalyptus* forest were very sharp, while others were diffuse, with evidence of rainforest trees invading the adjacent *Eucalyptus* forests. They noted that *Eucalyptus* did not regenerate beneath the ecotone and concluded that fire controlled the expansion of rainforest into adjacent *Eucalyptus* forests.

Two detailed studies have clearly demonstrated the importance of fire in controlling the position of a contrasting variety of rainforest types in the Australian subtropics. The abrupt boundaries between grassland and rainforest dominated by *Araucaria bidwillii* on the Bunya Mountains in southeast Queensland have puzzled ecologists for a long time (Herbert 1938; Webb 1964) (Figure 8.9). Fensham and Fairfax (1996) showed that these boundaries were not consistently related to soil depth, dry season soil moisture, altitude, rock cover, slope, landscape position or size of the grassland. Analysis of aerial photography over a 40-year period (1951–1991) demonstrated that both *Eucalyptus* forests and rainforest had invaded, and in some cases completely engulfed, grasslands. Fensham and Fairfax concurred with Herbert (1938) that fire is the probable

Figure 8.9
Naturally occurring grasslands between rainforest (on ridgeline) and Eucalyptus *forest in valley at the Bunya Mountains in southern Queensland. (Photograph: David Bowman.)*

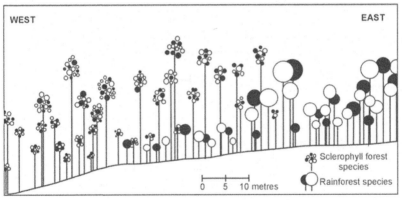

Figure 8.10
Profile diagram showing the expansion of subtropical rainforest into Eucalyptus saligna *forest on the northeast coast of NSW. Adapted from Smith and Guyer (1983).*

cause of the rainforest–grassland boundaries, and suggested that the source of fires were Aboriginal people. Smith and Guyer (1983) provided evidence that submontane vine forest had invaded *Eucalyptus saligna* forest on the northeast coast of NSW (Figure 8.10). They found that evidence of past fires (charcoal, charred logs and tree trunks and coppicing stems) was restricted to the

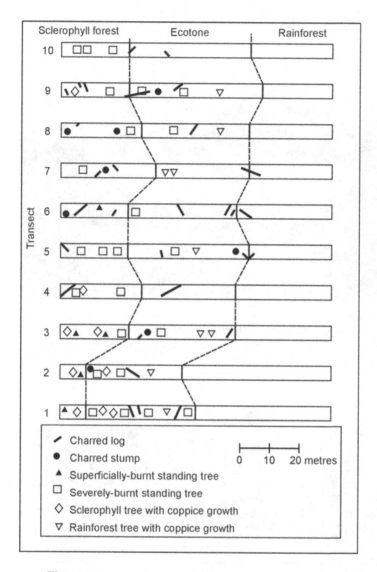

Figure 8.11

Distribution of evidence of fire across a Eucalyptus saligna–*subtropical rainforest boundary on the northeast coast of NSW. Adapted from Smith and Guyer (1983).*

Eucalyptus forest and ecotone (Figure 8.11), and concluded that fire limits the spatial extent of rainforest. Evidence of a recent fire corresponded with an abrupt boundary of young *Nothofagus moorei* trees, while evidence of a fire in the more distant past corresponded with an ecotone of medium-sized *N. moorei* trees (Figure 8.12).

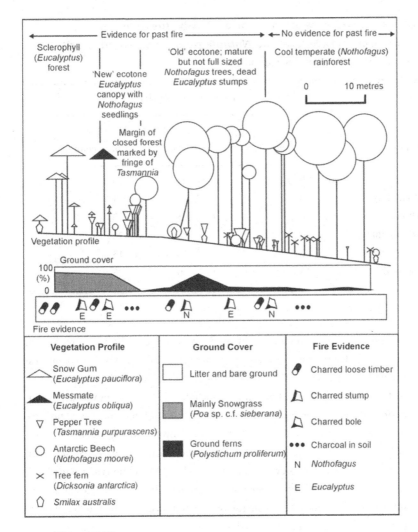

Figure 8.12

Profile diagram showing the expansion of Nothofagus moorei *rainforest into* Eucalyptus *forest on the Barrington Tops on the central coast of NSW. Distribution of evidence of fires across the boundary is also shown. Adapted from Smith and Guyer (1983).*

Temperate

Gilbert (1959) carried out a classic study of the role of fire in controlling temperate rainforests in the Florentine Valley in southern Tasmania. He showed that without periodic fire, forests dominated by *Eucalyptus regnans,* a tree which when fully grown is the tallest flowering plant on Earth, would be replaced by *Nothofagus cunninghamii* rainforest. Dense understoreys of rainforest trees of a

TABLE 8.11. Size class distribution of *Eucalyptus regnans*, *Atherosperma moschatum*, and *Nothofagus cunninghamii* at a site in the Styx Valley in western Tasmania

| Size class | Density of stems (ha^{-1}) | | |
	Eucalyptus regnans	Atherosperma moschatum	Nothofagus cunninghamii
30–130 cm stem height	0	435	15
0–10 cm dbh	0	2184	311
10–29 cm dbh	0	405	40
29–49 cm dbh	0	94	20
49–68 cm dbh	0	15	15
> 68 cm dbh	15	0	25

Adapted from Gilbert (1959).

range of size classes, and the lack of any *E. regnans* regeneration, was interpreted as clear evidence of a transition from mature *Eucalyptus* forest to rainforest (Table 8.11). Gilbert called such transitional forests 'mixed forests' and noted that they were spatially extensive. He concluded that *E. regnans* was dependent upon disturbance for regeneration, whereas the rainforest trees were not. He suggested landscape fire was the most common form of widespread disturbance, and this favoured regeneration of *Eucalyptus* because it triggered seed fall, removed dense understoreys and provided high light conditions at ground level. Dense stands of *Eucalyptus* regeneration stimulated by fire were also thought to suffer less insect and vertebrate herbivory. Gilbert speculated that, in the absence of fire eucalypts would be restricted to poor soils.

Mount (1979) took a more extreme view, arguing that in western Tasmania fire merely reinforced the restriction of rainforest to sites with well-drained soils. However, field research has demonstrated that edaphic factors are subordinate to the importance of fire protection in determining the distribution of rainforest. A superb example is provided by small quartzite islands in Bathurst Harbour which are covered with rainforest whilst on similar bedrock on the adjacent mainland, rainforest is restricted to large gullies (Brown and Podger 1982*b*). Nor is Mount's theory of the primacy of edaphic factors supported by historical records which clearly demonstrate that *Nothofagus cunninghamii* rainforests rapidly colonise unburnt *Eucalyptus* forests and grasslands across a range of soil types (Needham

Figure 8.13
Nothofagus cunninghamii *rainforest regenerating after a fire in northwestern Tasmania.*
(Photograph: David Bowman.)

1960; Ellis 1985). Ellis attributed this expansion in northeastern Tasmania to the cessation of Aboriginal burning.

Working in Victoria, Melick (1990*a*) suggested that were it not for landscape fires, warm temperate rainforests would spread out from sheltered gorges and gullies into wet *Eucalyptus* forests growing on more fire-prone sites. This conclusion is consistent with the findings of Gleadow and Ashton (1981), who documented the colonisation of *Eucalyptus* forests in central Victoria by the warm temperate rainforest tree *Pittosporum undulatum* some 200 km from its native range. Seeds from ornamental garden plants are thought to have been dispersed into the central Victorian *Eucalyptus* forests by the introduced European Blackbird (*Turdus merula*) (Gleadow 1982). *P. undulatum* is killed by high-intensity fires because it has thin bark and poorly protected basal buds, and has only been able to establish in *Eucalyptus* forest because of active fire suppression (Gleadow and Ashton 1981).

Many temperate rainforest trees have the capacity to recover following fire, and do so by a variety of mechanisms including vegetative re-growth and germination of seed (Howard 1973*a*; P. Barker 1991; Gibson and Brown 1991; Melick and Ashton 1991) (Figure 8.13). The critical factor controlling the distribution of rainforest trees appears to be their incapacity to survive recurrent

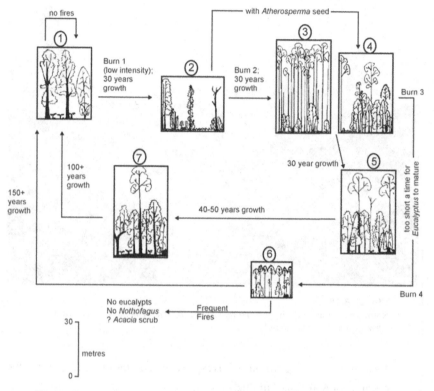

Figure 8.14

The effect of fire frequency on the regeneration of Nothofagus cunninghamii *forests in central Victoria. Codes for the vegetation types are as follows: 1: mature* Nothofagus *rainforest; 2:* Nothofagus *rainforest recovering from fire; 3:* Eucalyptus *forest with sparse* Nothofagus *understorey; 4:* Eucalyptus *forest with patchy* Atherosperma *and* Nothofagus *understorey; 5:* Eucalyptus *forest with dense* Atherosperma *and* Nothofagus *rainforest understorey; 6: Dense rainforest regeneration; 7: Rainforest with emergent* Eucalyptus. *Adapted from Howard (1973a).*

fires (Howard 1973a; Ashton and Frankenberg 1976). For instance, Howard noted that *Nothofagus cunninghamii* in central Victoria can recover from a single fire, but if a second fire occurs before canopy closure 40–50 years after the first fire, then the regenerating stand would be severely damaged. Each subsequent fire increases the risk of more intense fires that ultimately lead to the elimination of *N. cunninghamii* (Figure 8.14). Similarly, Gibson *et al.* (1991) concluded that the current regime of frequent fires has limited the distribution of the extremely long-lived endemic Tasmanian conifer *Lagarostrobos franklinii* (Figure 8.15), in spite of its ability to recover vegetatively from fire and dependence upon disturbance to initiate regeneration (Gibson and Brown 1991). Another endemic

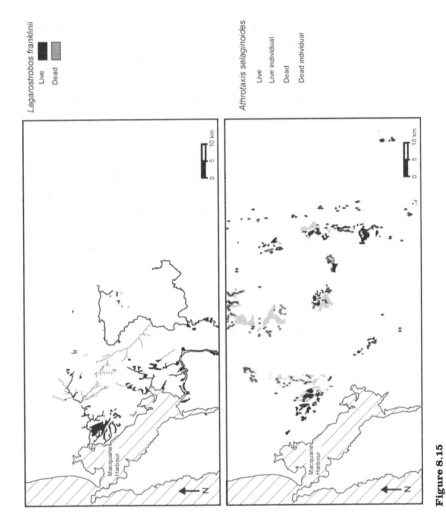

Figure 8.15

The distribution of living and dead Lagarostrobos franklinii and Athrotaxis selaginoides on the central west coast of Tasmania. Adapted from Anon. (1991).

Tasmanian rainforest conifer, *Athrotaxis selaginoides*, also regenerates following infrequent and localised catastrophic disturbances such as landslides and fire (Cullen 1987). However, *A. selaginoides* is unable to survive widespread, frequent and intense fires, having poor vegetative recovery abilities, infrequent seed production (about twice per decade), restricted seed dispersal and very slow growth. The high frequency of fires associated with European land management is thought to be responsible for the loss of about one third of *A. selaginoides* stands (Cullen 1987) (Figure 8.15).

Conclusion

Fire is an important factor in controlling rainforest boundaries across an enormous latitudinal and altitudinal gradient from the lowlands in the monsoon tropics to the mountains of Tasmania. It is remarkable that this generalisation holds given the fundamental influence that climate has on fire behaviour and frequency (Webb 1968). The extreme seasonality of the monsoon tropics promotes a high frequency of low-intensity fires in the winter dry season. In contrast, the temperate Australian forests are occasionally subject to summer weather conditions far more extreme than experienced in the seasonal tropics (Gill *et al.* 1996), with very high air temperatures, strong winds and low humidity that can contribute to fires of an astonishing intensity unknown elsewhere in Australia (Ashton 1981*a*). Webb (1968) suggested that only temperate rainforests protected by deep gorges could escape these infrequent and catastrophic fires. In the following chapter, I explore the effect of fire on nutrient cycles and ask whether it accounts for the difference in the fertility of soils across the rainforest boundary.

9

The fire theory II. Fire, nutrient cycling, and topography

Webb (1968) noted that although rainforests typically favour nutrient-rich or eutrophic soils, this correlation is not causal in the sense that rainforest trees have a physiological dependence upon high levels of soil nutrients. Rather, he argued that the occurrence of rainforest on eutrophic soils is more a consequence of the overarching effect of fire on the growth and survival of rainforest trees. Thus soil fertility is not merely determined by the underlying geology of a site but by fire history, nutrient cycling and topography. For instance, Webb argued that, in seasonally dry and geologically infertile environments, rainforest is restricted to topographically fire-protected sites, and that the absence of burning allows rainforest to enrich soils by nutrient cycling. The purpose of this chapter is to critically consider the role of nutrient cycling, fire and topography in creating differences in soil fertility between rainforest and non-rainforest vegetation.

Nutrient cycling and rainforest soil fertility

The view that differences in nutrient cycling contribute to the higher soil fertility of rainforest compared with adjacent non-rainforest is supported by the available data. For instance, Chandler and Lamb (1986) provide data on differences in nutrient use by rainforest and non-rainforest vegetation using the index of 'nutrient use efficiency' developed by Vitousek (1984). This is a measure of the total litter production for a given area of forest relative to the concentration of a given element in that litter (Mg of litter per ha/kg element per ha), which is an indication of the withdrawal of nutrients from the foliage prior to abscission. Despite considerable variation, Figure 9.1 illustrates that non-rainforest vegetation has higher nutrient use efficiencies than rainforest, and thus rainforest litter is richer in nutrients than non-rainforest litter. The finding of Webb *et al.* (1969)

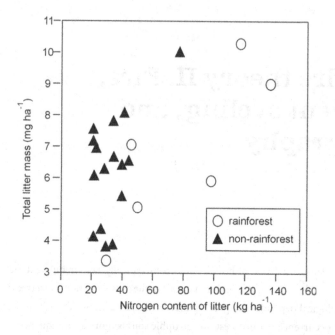

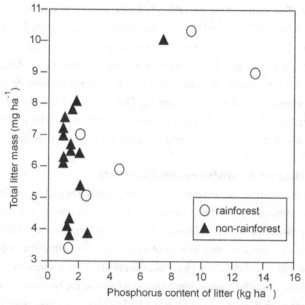

Figure 9.1

Relationship between total litter mass (produced over variable time periods) and the nitrogen and phosphorus content of that litter for a range of rainforest (circles) and non-rainforest (triangles) vegetation types. Values were collated from a range of published and unpublished sources by Chandler and Lamb (1986).

that, despite producing the same mass of litter, a subtropical rainforest returned twice the mass of nutrients than an adjacent *Eucalyptus pilularis* forest (Table 9.1), supports this conclusion.

However, differences in litter production between communities can strongly influence the total return of nutrients to the soil regardless of nutrient use efficiencies. For example, Webb *et al.* (1969) showed that adjacent *Ceratopetalum apetalum* and *Eucalyptus pilularis* forests returned a similar mass of nutrients in their forest litter fall over a 12-month period (Table 9.1) even though the *Eucalyptus pilularis* forest had greater nutrient use efficiencies than the *Ceratopetalum apetalum* forest. This is so because *Eucalyptus pilularis* produced a large volume of litter with relatively low concentrations of nutrients. Conversely, in the Northern Territory, rainforest soils are probably enriched by the greater mass of litter (Bowman and Wilson 1988), even though there is little difference in the nutrient contents of rainforest and savanna foliage and litter (see Table 5.8).

In summary, review of the sparse literature suggests that many rainforests enrich soil with litter fall, although by a variety of mechanisms.

Nutrients and rainforest succession

There is no doubt that soil nutrient capital is built up during primary rainforest successions. A good example of this is the development of relatively fertile soils over coastal sand dunes of Holocene age (Webb 1969; Fensham 1993). Some authors have suggested that the increased fertility of soils on sand dunes is transitory because of the occurrence of non-rainforest vegetation on infertile soils on old sand dunes. This pattern has been cited as evidence for the inevitable loss of nutrient capital during primary succession associated with over-leaching and subsequent podzolisation (Thompson and Bowman 1984; Fensham 1993). However, it is equally possible that nutrient loss could be related to changes in soil hydrology or fire or the interaction of these factors (Webb 1969).

Clear evidence that secondary succession to rainforest is necessarily accompanied by increased soil fertility is lacking. For instance, Ellis and Graley (1987) studied secondary successional sequences from *Eucalyptus* forest or grassland to *Nothofagus cunninghamii* rainforest on granitic soils in northeastern Tasmania. Ellis (1985) assumed that both the grassland and *Eucalyptus* forests are maintained by recurrent burning, and that the *Nothofagus* rainforest would replace theses communities in the absence of fire. Ellis and Graley found that there were few significant differences in the chemistry of surface soils in *Nothofagus* rainforest, grassland and *Eucalyptus* forest growing in a geologically uniform study area (Table 9.2). Exceptions were that the grassland had higher pH, total phosphorus,

TABLE 9.1. Concentrations of six nutrients in leaf litter fall in three different forest types in northeastern New South Wales

Forest type	Total leaf litter (Mg ha^{-1})	N (kg ha^{-1})	P (kg ha^{-1})	K (kg ha^{-1})	Ca (kg ha^{-1})	Al (kg ha^{-1})
Subtropical rainforest	5.9	97	4.7	30.5	89.7	8.3
Ceratopetalum apetalum rainforest	3.4	30	1.3	7.2	30.6	21.3
Eucalyptus pilularis forest	6.5	41	1.3	8.3	24.9	6.9

Adapted from Webb *et al.* (1969).

TABLE 9.2. Chemical characteristics from the soil surface (0–10 cm) of three vegetation types growing on granitic soils in northeastern Tasmania. Ellis (1985) has argued that the grassland and *Eucalyptus* forest are seral communities in a pyric succession toward *Nothofagus cunninghamii* rainforest

	Vegetation type		
Chemical variable	*Poa* grassland	Grassy *Eucalyptus* forest	*Nothofagus cunninghamii* rainforest
pH	5.4 a	5.31 a	4.52 b
Loss on ignition (g $100\,g^{-1}$)	18.2 a	16.4 a	19.8 a
N (g $100\,g^{-1}$)	0.48 a	0.32 a	0.41 a
P (g $100\,g^{-1}$)	0.070 a	0.060 ab	0.045 b
Ca (g $100\,g^{-1}$)	0.39 a	0.31 a	0.27 a
Mg (g $100\,g^{-1}$)	0.12 a	0.14 a	0.11 a
K (g $100\,g^{-1}$)	2.7 a	2.6 a	2.6 a
Ca (Meq $100\,g^{-1}$)	2.69 a	3.27 a	1.87 a
Mg (Meq $100\,g^{-1}$)	0.89 a	1.35 a	1.02 a
K (Meq $100\,g^{-1}$)	0.45 a	0.39 a	0.34 a
CEC (Meq $100\,g^{-1}$)	25.0 a	28.0 a	26.8 a
NH_4-N (%)	N.A.	0.152 b	0.231 b
NO_3-N (%)	3.38 a	0.042 ab	0.020 b

Letters denote mean values not significantly different at $P = 0.05$.
Adapted from Ellis and Graley (1987).

total nitrogen, and nitrate, and lower ammonium. The *Eucalyptus* forest was found only to have higher pH than the rainforest. Although these data do not support the idea that secondary succession results in increased soil fertility, it is likely that the nutrient capital increases with succession towards rainforest, but that the nutrients are stored in the biomass rather than in the soil. Ellis and Graley's data do not support the strict interpretation of Jackson's (1968) opinion that in southwest Tasmania a sedgeland with infertile organic soils that remained unburnt for 800 years would develop into a rainforest with fertile organic soils.

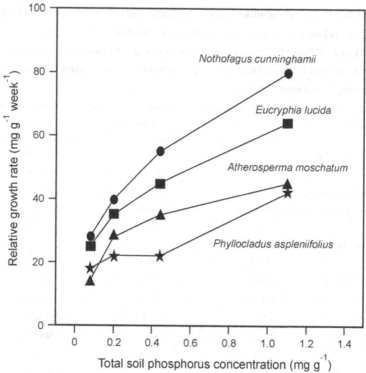

Figure 9.2

Relationship between seedling growth rate of four rainforest tree species and the total phosphorus concentration in the surface soil (0–20 cm) collected from rainforest on four different parent materials. The seedlings collected from the field were grown in pots containing the forest soils for 196 days. Adapted from Read (1995).

Indeed, field surveys in southwest Tasmania led Bowman *et al.* (1986) to suggest that an extraordinarily long fire-free interval (i.e. millennia) would be required for sedgeland to accumulate sufficient nutrients to produce soils as fertile as those found supporting rainforest vegetation. Furthermore, it is unlikely that succession to rainforest would happen as rapidly as predicted by Jackson given that rainforest seedling growth is controlled by soil fertility (Howard 1973*a*; Read 1995) (Figure 9.2).

Clearly more research is required to determine the amount of nutrients accumulated by primary and secondary rainforest succession. The sparse available evidence does not support the generalisation that secondary succession necessarily results in increased surface soil fertility, and where it does, nutrient accumulation appears to require millennia, as is the case for primary successions.

Nutrient loss following fire

A number of authors have stressed the role of fire in creating differences in soil fertility between rainforest and non-rainforest vegetation (Jackson 1968; Stocker 1969), but there have been no published studies that have documented nutrient losses following fire in Australian rainforest. The only studies that shed light on the question relate to the consequences of the combustion of logging debris in mixed *Nothofagus* rainforest – *Eucalyptus* forest in Tasmania. It is difficult to draw parallels between wildfire and clear felling and slash burning because of differences in the moisture content and arrangement of fine fuels (grasses and twigs), duration of fires and soil disturbance associated with logging (Raison 1980). Nonetheless, these applied ecological studies give an insight into the possible importance of fire in creating differences, and are therefore briefly reviewed.

Harwood and Jackson (1975) found that, after a slash fire in a mixed *Nothofagus* rainforest–*Eucalyptus* forest in Tasmania, 10 kg ha^{-1} of phosphorus, 51 kg ha^{-1} potassium, 100 kg ha^{-1} calcium, and 37 kg ha^{-1} magnesium were lost to the atmosphere, equal to about 18%, 17%, 12% and 29% respectively of the nutrient content of the fuel consumed (Figure 9.3). Whether this atmospheric loss is of any long-term significance is debated and requires further research. Ellis and Graley (1983) suggest that these losses would be replaced by nutrients contained in rainwater within 15 to 20 years.

Ellis and Graley (1983) studied the immediate effect of a fire on *Nothofagus* rainforest–*Eucalyptus* forest soils derived from dolerite, an igneous rock type of moderate fertility. The fire in logging debris caused the loss of 35% of the carbon and 39% of the nitrogen, but led to an increase in the total concentrations of phosphorus (5%), calcium (61%), magnesium (29%) and potassium (28%). Burning increased pH and soil exchangeable calcium, magnesium, potassium, extractable phosphorus and mineral nitrogen. Ellis *et al.* (1982) compared uncut 'control' *Eucalyptus–Nothofagus* rainforest soils with soils from similar forests that had been logged and burnt up to 4 years previously. They concluded that the pulse of available nutrients following the combustion of logging debris remained in the upper 10 cm of soil for at least 4 years. They found no evidence of significant nutrient loss by leaching of nutrients into the subsoil, presumably because rapidly regenerating trees progressively take up the available nutrients released by burning. Lost carbon and nitrogen had not been replaced four years after burning, although there were indications that these elements may be gradually replaced by litter fall from the regenerating forest.

A hotly debated aspect of the impact of logging and burning on forest nutrient

Figure 9.3
Smoke from a forest fire transports nutrients from the site. The long-term ecological significance of this nutrient loss remains unclear. (Photograph: David Bowman.)

capital has been the appropriate time interval between these disturbances. Based on a number of assumptions about losses and inputs, Raison (1980) concluded that the replacement time of lost phosphorus following logging and burning would be about 150 years, but using different assumptions, Neilsen and Ellis (1981) suggested that phosphorus would be replaced within 40 years. In the case of burnt but unlogged mixed *Eucalyptus–Nothofagus* rainforest stands (i.e. where biomass is not exported from the stand), the calculations of both Raison (1980) and Neilsen and Ellis (1981) suggested that phosphorus would be replaced within 100 years and 25 years respectively. These studies suggested that a regime of frequent burning (i.e. more than one fire every century) would need to be maintained for very long periods to cause substantial declines in soil nutrients on moderately fertile substrates. This conclusion is consistent with the view of Ashton (1976a). He found that the concentration of phosphorus in the surface soil of moist *Eucalyptus* forest on a sheltered slope was similar to that of the *Nothofagus* forest in the floor of the granite basin. However, both the rainforest and moist *Eucalyptus* forest had higher concentrations of surface soil phosphorus that an adjacent dry *Eucalyptus* forest on a hot north-facing slope (see Chapter 5). Ashton suggested that differences in fire frequency 'sustained for many thousand of years' may be 'ultimately important' in accounting for the phosphorus concentrations on opposite sides of a geologically uniform granite basin in central Victoria. He suggested that recurrent fires in the dry *Eucalyptus* forest may have lowered soil phosphorus concentrations compared to the cooler, more mesic and less frequently burnt south-facing slope. An alternative, though not mutually exclusive hypothesis considered by Ashton was that past phases of aridity and erosion may have created the contrast in soil fertility. These hypotheses are consistent with Macphail's (1980, 1991) argument, based on Tasmanian palynological data, that in addition to the effects of drier climates, the shift towards non-rainforest vegetation in the late Holocene was associated with declining soil fertility caused by over-leaching and nutrient losses following wildfire.

In summary, the scant evidence suggests that very long periods of frequent burning would be required to cause significant chemical differences between rainforest and non-rainforest vegetation. The impact of nutrient loss following fire is assumed to be of much greater significance on oligotrophic substrates such as quartzite, especially if organic soils are consumed by ground fires (Jackson 1968; Bowman and Jackson 1981; Brown and Podger 1982*b*; Hill 1982). On such infertile substrates it is widely assumed that nearly all the nutrient capital has been gradually accumulated from atmospheric inputs.

Topography and soil fertility

Rainforests are often restricted to steep gullies. Although erosional processes may result in the enrichment of soils in gullies (Beadle 1962*a*: Webb 1968), most authors have concluded that the relationship between gullies and rainforest is a result of fire protection and the more humid microclimate rather than soil fertility. This point can be illustrated by considering the relationship between topography, soil fertility and *Ceratopetalum apetalum* rainforests that are typically restricted to sandstone gullies in the Sydney region (Figure 9.4). Beadle (1962*a*) suggested that slope wash of nutrients accounted for surface soil phosphorus concentrations in the rainforest gullies that were three times greater than that of their sandstone parent material. Subsequent work does not support this interpretation. Clements (1983) demonstrated that soils in sandstone gullies that

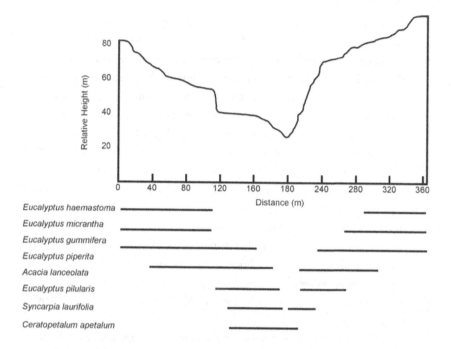

Figure 9.4

Profile diagram showing the restriction of Ceratopetalum apetalum *to the floor of a sandstone gully in the Sydney region. Adapted from Pidgeon (1940).*

often supported *C. apetalum* rainforest had a lower mean concentration of phosphorus than soils in other topographic positions (Figures 4.3 and 4.4). The restriction of *C. apetalum* rainforest to sandstone gullies is most likely explained by other factors such as soil moisture or fire protection (McLuckie and Petrie 1927; Pidgeon 1937, 1940; Baur 1957; Florence 1964; Le Brocque and Buckney 1994), although the drought tolerance of many rainforest trees does not support the view that rainforests are restricted to gullies because of favourable moisture supply (see Chapter 6).

Conclusion

In this chapter, I have shown that patterns of nutrient cycling vary between rainforest types. The sparse available evidence suggests that very long periods of recurrent fires are required to deplete the nutrient capital of moderately fertile rainforest soils. Rapid nutrient losses may occur in oligotrophic environments, particularly if substantial amounts of soil organic matter are consumed by burning. Equally, the accumulation of nutrients by rainforests in oligotrophic environments may be very slow. Jackson developed a theory that emphasised the interactive relationships between fire history, soil fertility and vegetation, including rainforest in southwest Tasmania. This model is critically discussed in the following chapter.

10

The fire theory III. Fire frequency, succession, and ecological drift

In Chapter 8 I summarised a large body of field evidence, collected throughout the geographic range of Australian rainforest, that strongly suggested that fire plays a major role in controlling rainforest boundaries. However, I also showed that the relationship between fire and rainforest distribution is not simple. Rainforest trees are best differentiated from non-rainforest tree species by their inability to survive *recurrent* fires; most rainforest tree species can regenerate after a *single* fire. Thus, the *frequency* of fires is critical in controlling rainforest. In the previous chapter, I showed that differences in soil fertility across rainforest boundaries reflect the cumulative effect of fire history. In an influential paper published in 1968, Jackson emphasised the role of fire frequency in determining soil fertility and vegetation types in western Tasmania. Specifically, he developed the 'ecological drift' theory that emphasised the importance of variation in fire frequency in determining the distribution of vegetation types, including rainforest, in western Tasmania. Has Jackson found the key to understanding the distribution of Australian rainforest? In this chapter, I critically review Jackson's theory.

Rainforest, succession and the ecological drift model

Jackson (1968) set out to answer a simple question: why should rainforest be spatially restricted in western Tasmania? With its cool, mid-latitude oceanic climate and an annual rainfall in excess of 2000 mm distributed throughout the year, western Tasmania would seem ideal for cool temperate rainforest. However, rainforest is of limited extent and treeless sedgelands dominated by *Gymnoschoenus sphaerocephalus* (commonly known as button-grass) are widespread (Figure 10.1). Jackson considered this situation anomalous, noting that 'there would appear to be no aspect, soil-type, or edaphic situation within this region

Figure 10.1
The large area of treeless sedgelands is a characteristic feature of southwest Tasmania. Given the region's high rainfall, why do these areas not support temperate rainforest? (Photograph: David Bowman.)

which cannot be occupied by cold temperate rainforest'. To explain the anomaly, Jackson built on the empirical observations of Gilbert (1959), who had studied the relationship between fire and tall *Eucalyptus* forests and rainforest in western Tasmania, as well as his own substantial, though largely unpublished, field experience. Fundamentally, he believed that all non-rainforest communities in western Tasmania would ultimately develop into self-perpetuating 'climax' rain-forests given a complete absence of fire because he assumed that rainforest tree species such as *Nothofagus cunninghamii* were the only plants that could regenerate in the absence of fire.

Jackson argued that vegetation, fire frequency and soil fertility interacted in a complex system of feedback loops (Figure 10.2). He developed a probabilistic model that is analogous to the concept of 'genetic drift' (Figure 10.3) and is therefore known as the 'ecological drift' model. The crux of the ecological drift model is the belief that the existence of vegetation types are held in a 'delicate balance' in response to varying time intervals between successive fires. Although he believed there is a strong central tendency in the probability distributions of the fire-free interval for each vegetation type, he suggested that random deviations from the mean fire-free intervals would cause the 'drift' of one vegetation type to another vegetation type. Such drift resulted in a corresponding change in

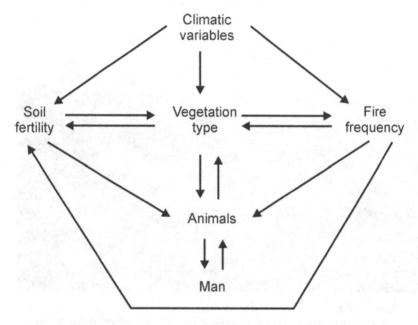

Figure 10.2
Hypothesised ecological interactions that determine vegetation types in western Tasmania. Adapted from Jackson (1968).

the probability distribution of various fire-free intervals. For example, if a *Eucalyptus* forest recovering from a fire in the very recent past were burnt again, then the regenerating forest would be transformed into a treeless sedgeland because the recently burnt forest would lack reproductively mature trees to provide the seeds needed to initiate the regeneration of the forest. The potential role of seeds stored in the soil or the long-distance dispersal of seeds was not considered. Associated with this 'drift' to another vegetation type was a corresponding change in the probability of fire. Sedgelands were assumed to have a far higher probability of short fire-free intervals than the forests, and the model predicted that sedgelands could rarely become forested because of the improbability of a sufficiently long fire-free interval to permit the establishment and maturation of trees (Figure 10.3). A strict interpretation of Jackson's model shows that a sedgeland would be converted into rainforest if it remained unburnt for 850 years (Henderson and Wilkins 1975). Although the drift model was based on the thought that vegetation patterns were dynamic, it also predicted that the overall proportion of the landscape covered by different types of vegetation was in equilibrium. This is so because the succession was assumed to be controlled by fire-free intervals with a low probability of occurrence.

The applied mathematicians Henderson and Wilkins (1975) used a transition probability matrix model, sometimes known as a Markov model, to describe the ecological drift model mathematically. They found that the distribution of rainforest was very sensitive to changes in fire frequency (Figure 10.4). When they used Jackson's estimates of the probability of different fire-free intervals, their model was broadly in agreement with observed vegetation patterns at local and river catchment scales in southwest Tasmania (Brown and Podger 1982a) (Table 10.1). This broad agreement is remarkable given some assumptions made by Jackson and by Henderson and Wilkins. These assumptions were as follows:

(i)   discrete vegetation types exist;
(ii)   it is possible to determine the probability of various fire-free intervals;
(iii) fire occurs randomly in space and time;
(iv)   fire acts uniformly in the landscape; and
(v)   edaphic conditions do not influence the rate of succession.

Each of these assumptions is discussed critically below.

Discrete vegetation types

Philosophically, the 'ecological drift' model embraced the now unpopular idea of 'classical' secondary succession developed by the American ecologist Clements (Noble and Slatyer 1980). The Clementsian view is that, without disturbance, one vegetation type known as the 'climatic climax' would dominate a particular climatic zone. The climax was the terminal stage of both primary and secondary successions. Primary succession concerns the establishment of vegetation on previously barren land surfaces and does not concern us here. Secondary succession concerns the re-establishment of vegetation after a disturbance such as a forest fire. Clements believed that recovery of the 'climatic climax' followed a highly predictable series of vegetation types called 'seral stages'. The dominant plants in the previous seral stage are believed to create environmental conditions that are essential for the establishment of plants that dominate the subsequent seral stage and so on until the climax stage is reached. Connell and Slatyer (1977) described this orderly chain of replacement as 'facultative succession'.

The 'ecological drift' model was based on the assumption that the vegetation of western Tasmania comprises five seral stages (sedgeland, shrubland, wet scrub, wet sclerophyll forest and mixed forest) that lead to the rainforest 'climax'. The belief that the successional vegetation sequence is made up of discrete communities is a fundamental flaw of the 'ecological drift' model. Field evidence shows that the vegetation in western Tasmania forms continua in both space and time.

Brown and Podger (1982b) and Brown et al. (1984) clearly demonstrated that there is a floristic continuum between sedgeland and rainforest, and that this continuum is correlated to time since the last fire (Figures 10.5 and 10.6).

Noble and Slatyer (1980, 1981) sought to overcome this unrealistic assumption by developing the 'vital attributes' model. Vital attributes are life-history parameters of individual taxa including growth rate, age of maturity and age of senescence, dispersal and persistence of propagules, and dependence of seedling establishment on disturbance. They modelled the successional paths of western Tasmania forests to different fire-free intervals by studying the 'vital attributes' of four western Tasmania tree species (*Acacia dealbata*, *Eucalyptus regnans*, *Nothofagus cunninghamii* and *Atherosperma moschatum*) (Figure 10.7). These four woody species were thought to be characteristic of distinct successional stages from *Eucalyptus* forest to rainforest.

Moore and Noble (1990) refined the vital attributes model by considering the possible effect of competition between plant species, and Noble and Gitay (1996) modified the model to account for spatial variation in vegetation dynamics. However, as will be discussed below, both these models and the ecological drift

---

Figure 10.3

Flow diagram showing the assumed probability distributions for fire-free intervals in six vegetation types in western Tasmania. Shifts from one vegetation type to another are assumed to be governed by specific fire-free intervals that arise due to chance. The probabilities and their frequency distributions are indicative only.

Solid arrows described the replacement of woody vegetation types (rainforest, mixed Eucalyptus–Nothofagus *forest, wet sclerophyll forest, wet scrub) with sedge-heath vegetation, or the replacement of sedge-heath with sedgeland. Such transitions are thought to come about because two fires occur in quick succession, killing regenerating woody plants before they are sexually mature. The probability of this happening is shown as dark shading for each vegetation type.*

Dotted arrows described the replacement of mixed Eucalyptus–Nothofagus *forest by wet sclerophyll forest, wet sclerophyll forest by wet scrub and wet scrub by sedge-heath. These transitions are thought to come about because the interval between two successive fires is insufficiently long to allow the sexually maturity of the dominant plants in vegetation (e.g.* Eucalyptus *in wet sclerophyll forests) but sufficiently long for the maturity of plants in the previous seral stage (e.g.* Melaleuca *in wet scrub). The probability of this happening is shown in dotted shading.*

Dashed arrows describe the replacement of sedge-heath by wet scrub, wet scrub by wet sclerophyll forest, wet sclerophyll forest by mixed Eucalyptus–Nothofagus *forest and mixed* Eucalyptus–Nothofagus *forest by rainforest. These transitions are thought to come about because the interval between two successive fires is sufficiently long to allow the establishment and sexual maturity of the long-lived plants that would otherwise be eliminated by shorter fire-free intervals. The probability of this happening is shown by barred shading.*

Unshaded areas of each graph denote fire-free intervals that would maintain each of the given vegetation types. The operational details of the model are critically discussed in the text. Adapted from Jackson (1968).

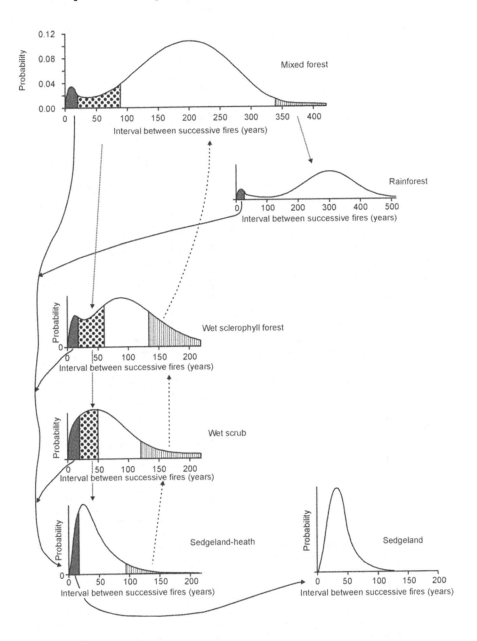

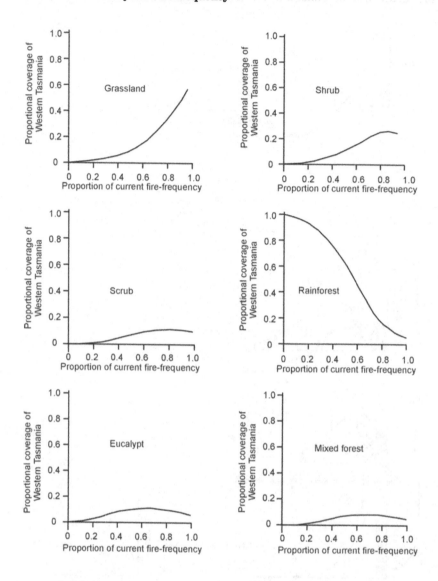

Figure 10.4

Predicted proportion of western Tasmania covered by six vegetation types in response to variations in the frequency of fires according to the mathematical model of Henderson and Wilkins (1975). A fire frequency of 1 is assumed to reflect the current fire frequency based on the estimates of fire-free intervals provided by Jackson's (1968) diagrammatic model (Figure 10.3). Adapted from Henderson and Wilkins (1975) and reproduced from Search *(now published as* Australasian Science*).*

TABLE 10.1. Coverage of six vegetation types in western Tasmania according to the predictions of Henderson and Wilkins' (1975) mathematical model and the observations of Brown and Podger (1982a)

Brown and Podger (1982a) did not record observations on the coverage of the 'shrub' vegetation type of Henderson and Wilkins (1975). Here it is assumed that Brown and Podger's (1982a) 'scrub' and 'sclerophyll' vegetation types are equivalent to the 'shrub' and 'eucalypt' + 'scrub' vegetation types of Henderson and Wilkins (1975) respectively. Henderson and Wilkins predicted that 10% of the vegetation would be in an undefined 'non-mature state' and this proportion is not listed in the table.

Proportion of coverage (%)	Vegetation types defined by Henderson and Wilkins (1975)					
	Rainforest	Mixed forest	Eucalypt	Scrub	Shrub	Grassland
Predicted	2.5	1.5	2.5	7	23.5	53
Observed in 2 km² at Bathurst Harbour	7	10	12		15	56
Mean and range for 12 catchments in southwest Tasmania	10 (2–25)	12 (4–24)	14 (2–29)		27 (16–43)	37 (10–62)
Overall, southwest Tasmania	11	12	13		25	39

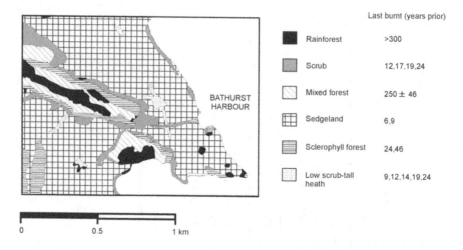

		Last burnt (years prior)
■	Rainforest	>300
▨	Scrub	12,17,19,24
▨	Mixed forest	250 ± 46
▦	Sedgeland	6,9
▤	Sclerophyll forest	24,46
▫	Low scrub-tall heath	9,12,14,19,24

Figure 10.5
Vegetation map of a small area beside Bathurst Harbour in southwest Tasmania. The time since the last fire in various stands of each vegetation type is shown. Adapted from Brown and Podger (1982b).

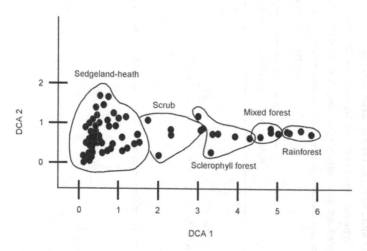

Figure 10.6
Ordination of floristic quadrat data collected from different vegetation types at Bathurst Harbour. The first axis of the ordination (DCA1) is strongly correlated with the time since the last fire. Adapted from Brown et al. (1984).

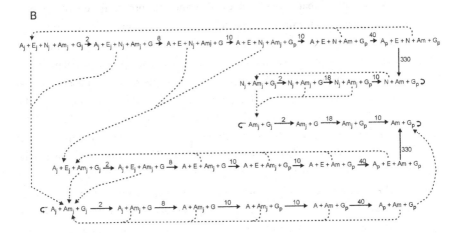

Figure 10.7

Model of the response of five common western Tasmanian taxa (four western Tasmanian tree species (Acacia dealbata, Eucalyptus regnans, Nothofagus cunninghamii *and* Atherospermum moschatum) *and graminoids) to various time periods between fire disturbance.*

(A) Each of the five taxa is classified into functional groups according to the system of Noble and Slatyer (1980). In this system 'S' indicates persistence of long-lived propagules, 'C' indicates persistence of short-lived propagules, 'T' indicates that propagules can establish in the absence of fire, 'I' indicates that propagules can establish only after fire.

The chronology of life stages following a fire regeneration event is also indicated for each of the taxa where: 'M' is the age of reproductive maturity, 'I' is the age when it becomes senescent and 'e' is when it becomes locally extinct.

(B) Pathways of successional stand transformation following various combinations of fire-free intervals in a hypothetical stand of vegetation comprised of five common western Tasmanian taxa. Solid lines denote stand transformation following fire-free disturbance. Dashed lines denote stand transformations following fire disturbance. The subscript 'j' denotes an immature population of a given taxon and the subscript 'p' denotes the occurrence of a propagule population. The integers above each solid line indicate the time (in years) elapsed between stand transformations. Adapted from Moore and Noble (1990), by permission of the publisher Academic Press.

model unrealistically assumed that the intensity of fire is constant. Noble and Slatyer (1980) acknowledged that variation in the intensity of fire may change vital attributes such as the ability of mature and immature plants to survive the fire, and thus change the predictions of their model. A further weakness with the approach is that 'vital attribute' data are unavailable for most Australian trees. Furthermore, it is difficult to realistically determine the effect of edaphic and climatic variation on vital attributes such as growth rate, age of maturity and variation in seed production.

Quantification of fire-free intervals

Jackson believed that it was possible to determine the probability distributions of fire-free intervals for the six vegetation types by estimating 'plant age on a large number of parallel sites for each vegetation type'. Nonetheless he only provided indicative probability distributions for each vegetation type and actual probability values for one vegetation type (mixed forest) (Figure 10.3). In the case of the three non-forest communities (sedgeland, shrubland, wet scrub), he assumed the probability distributions were skewed and unimodal reflecting the high risk of frequent fires. In the case of the three forest communities (rainforest, mixed forest and wet sclerophyll forest), the probability distributions are assumed to be bimodal (Figure 10.3). Bimodal distributions were justified because Jackson believed there were different fire risks in regenerating forest than older, more mature forests. The assumed relationship between the probability of different fire-free intervals and the structural development of a 'mixed forest' is illustrated in Figure 10.8.

Henderson and Wilkins (1975) divided the successional sequence up into 12 vegetation categories (Figure 10.9). With the exception of Jackson's sedgeland and mixed forest, which they treated as special cases, vegetation types were considered to exist in either mature or immature sub-states. The mature sub-state denoted vegetation that had remained unburnt for a long enough period to allow the maturation of the dominant plants. Should this sub-state be burnt, seeds are available for its re-establishment. The immature sub-state denoted vegetation recovering from a fire in the recent past such that it is composed entirely of immature plants. Should fire occur at this stage, the vegetation would be destroyed because of the absence of any seeds. In the case of the sedgeland, which is the terminal 'disclimax' community, it was assumed that the community is always at risk of being burnt and thus there is only one state. In the case of the mixed forest there are three sub-states reflecting: (a) the possibility of conversion of mixed forests to treeless vegetation; (b) the possibility of

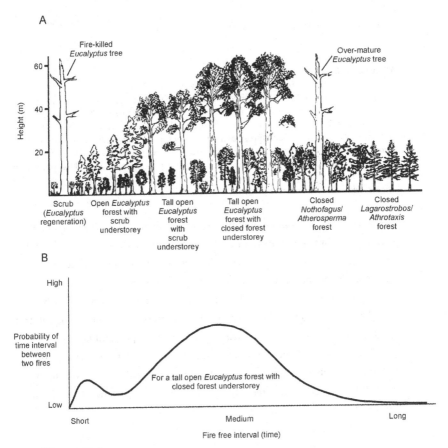

Figure 10.8

(A) Assumed structural development of a mature mixed Eucalyptus–Nothofagus *forest following destruction by wildfire. Fire in mature mixed* Eucalyptus–Nothofagus *forest is assumed to destroy the forest trees but stimulate* Eucalyptus *regeneration that eventually forms an even-aged tall open forest. Given a long fire-free interval, rainforest develops and prevents the further regeneration of* Eucalyptus *trees. Eventually the* Eucalyptus *trees die from old age. (B) Probability distribution of fire-free intervals for a mixed* Eucalyptus–Nothofagus *forest regenerating after fire. The openness of young regenerating stages is assumed to elevate the risk of fire and thus increase the probability of a short interval before the next fire. Mature stands are thought to be less fire-prone and thus have a central tendency for their fire-free interval distributions. Medium fire-free intervals favour the regeneration of mixed* Eucalyptus–Nothofagus *forest by re-initiating the succession described in diagram A. Long fire-free intervals result in the death of* Eucalyptus *canopy trees and their complete replacement with rainforest trees. The contrasting fire risk in regenerating and mature* Eucalyptus *forests produces the bimodal probability distribution of fire-free intervals. Adapted from Bowman (1986) with kind permission from Kluwer Academic Publishers.*

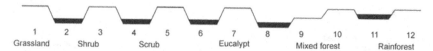

Figure 10.9

The 12 fire-risk states defined in Henderson and Wilkins' (1975) model of the forest dynamics in western Tasmania. The 'danger periods' are shaded and denote open regenerating plant communities. There is only one state for the grassland and three states for the mixed forest, the latter reflecting (i) the possibility of conversion of mixed forests to treeless vegetation, (ii) the possibility of perpetuating Eucalyptus domination by a fire occurring after the Eucalyptus trees had matured but before the rainforest trees had matured, and (iii) the possibility of conversion of mixed forest to rainforest following the senescence of the Eucalyptus trees and the maturity of the rainforest trees. Adapted from Henderson and Wilkins (1975) and reproduced from Search (now published as Australasian Science).

perpetuating *Eucalyptus* domination by a fire occurring after the *Eucalyptus* trees had matured but before the rainforest trees had matured; and (c) the possibility of conversion of mixed forest to rainforest following the senescence of the *Eucalyptus* trees and the maturity of the rainforest trees.

The lack of specific probabilities for various fire-free intervals clearly frustrated Henderson and Wilkins' mathematical treatment of the ecological drift model. They overcome this obstacle by estimating the probabilities by assuming that fires occurred randomly through time and thus that the occurrence of fires could be described by a Poisson distribution. Given a Poisson distribution, the probability of a given fire-free interval (*t*) for a given vegetation type (i) can be calculated by the equation:

$$e^{-\lambda_i t} \tag{10.1}$$

where λ_i is the frequency of a fire for a given vegetation type.

This term can be derived from Jackson's diagram (Figure 10.3) because $1/\lambda_i$ is the average time between fires (the fire-free interval) for the vegetation type i. They assumed that the rate of succession (i.e. time elapsed before a given vegetation state (D_i) moves to another state (D_j)) is constant and specific for each of their 12 vegetation types (Figure 10.9), and they estimated these values from Jackson's original figure (Figure 10.3). Again, assuming the occurrence of fire is random, the probability of a transition from one of the 12 vegetation types to the next vegetation type ($f_{(i)}$) is found by modifying equation 10.1 to:

$$f_{(i)} = e^{-\lambda_i D_i} \tag{10.2}$$

The reciprocal of this transitional probability, $1 - f_{(i)}$, returns the vegetation type to a 'lower' vegetation state in the successional sequence. Henderson and Wilkins illustrated this point by taking the case of the mature scrub community. If this

TABLE 10.2. Estimates of the average fire-free interval, frequency of a
given fire-free interval and the transition time between the 12 vegetation
types (i) used in Henderson and Wilkins' (1975) model

Vegetation type (state) (i)	Average fire-free interval $1/\lambda_i$ (years)	Fire frequency λ_i (years^{-1})	Transition time between vegetation states D_i (years)
Sedgeland (1)	25	0.04	50
Regeneration shrub (2)	25	0.04	20
Mature shrub (3)	50	0.02	100
Regeneration scrub (4)	30	0.003	20
Mature scrub (5)	60	0.017	120
Regeneration *Eucalyptus* forest (6)	40	0.025	20
Mature *Eucalyptus* forest (7)	90	0.011	150
Regeneration mixed forest (8)	50	0.02	20
Mixed forest (9)	90	0.011	20
Mature mixed forest (10)	200	0.005	250
Regeneration rainforest (11)	90	0.011	40
Mature rainforest (12)	300	0.003	∞

vegetation type (state D_5) remains unburnt for at least 120 years then it is
transformed into a mature *Eucalyptus* forest (state D_7). Conversely, if it is burnt
in less than 120 years then it is transformed into immature scrub (state D_4) that is
at risk of complete destruction and conversion to sedgeland (state D_1). The
values of $1/\lambda_i$, λ_i and D_i are listed in Table 10.2. The matrix of the transition
probabilities between the 12 vegetation states used in the Henderson and Wilkins
model are listed in Table 10.3.

Despite Henderson and Wilkins' ingenious method of estimating probabilities
of various fire-free intervals, it is actually impossible to estimate these probabilities. This is because the succession from sedgeland to rainforest is not made up of
discrete seral communities but rather they form a structural and floristic continuum (Figures 10.5 and 10.6). The only way to quantify the probabilities of
fire-free intervals is by establishing longitudinal studies or carrying out historical
reconstructions. However, such studies could only be related to particular sites,
not particular vegetation types.

TABLE 10.3. Matrix showing the transitions between the 12 vegetation states defined by Henderson and Wilkins (1975) in their model of the vegetation dynamics in western Tasmania (see Table 10.2).

Current state	Next vegetation state											
	1	2	3	4	5	6	7	8	9	10	11	12
1	0.86	—	0.14	—	—	—	—	—	—	—	—	—
2	0.55	—	0.45	—	—	—	—	—	—	—	—	—
3	—	0.86	—	—	0.14	—	—	—	—	—	—	—
4	0.06	—	—	—	0.94	—	—	—	—	—	—	—
5	—	—	—	0.87	—	—	0.13	—	—	—	—	—
6	0.39	—	—	—	—	—	0.61	—	—	—	—	—
7	—	—	—	—	—	0.81	—	—	—	0.19	—	—
8	0.33	—	—	—	—	—	—	—	0.67	—	—	—
9	—	—	—	—	—	0.20	—	—	—	0.80	—	—
10	—	—	—	—	—	—	—	0.71	—	—	—	0.29
11	0.36	—	—	—	—	—	—	—	—	—	—	0.64
12	—	—	—	—	—	—	—	—	—	—	1.0	—

Adapted from Noble and Slatyer (1981), by permission of the Australian Academy of Science.

   Emphasis on specific sites rather than generalised vegetation types is incompatible with the ecological drift model. Although the interaction between fire frequency, soil fertility and vegetation types was central to Jackson's (1968) argument concerning the ecological determinants of vegetation distribution in western Tasmania (Figure 10.2), he did not include this interaction in his ecological drift model. Field evidence clearly demonstrates that site characteristics such as topographic position influence the risk of fire, as Jackson himself demonstrated. At the local scale, he showed that ridge tops and slopes facing prevailing winds are more prone to fire than sheltered slopes and valleys (Figure 10.10). Similarly, Ashton (1981a) and P. Barker (1991) have demonstrated that topographic position influences fire history in Victoria, with rainforest occurring in infrequently burnt gullies and southeast facing slopes, while ridges and north-facing slopes are more frequently burnt (Figure 10.11).
   Mount (1964, 1979, 1982) proposed a controversial alternative model to the

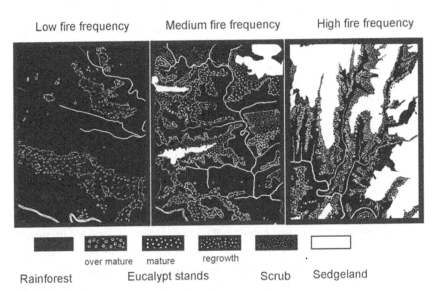

Figure 10.10
The effect of fire frequencies and topography on vegetation coverage in western Tasmania. With increasing fire frequency, rainforest is progressively replaced by Eucalyptus *forest, then by scrub and ultimately sedgeland. Topography is indicated by major creeklines. Adapted from Jackson (1968).*

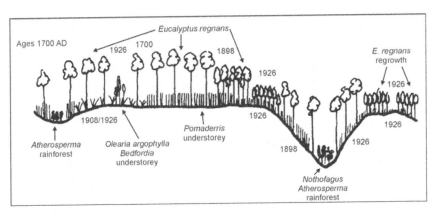

Figure 10.11
Relationship between time since last fire, topography and understorey type in Eucalyptus regnans *forests at Wallaby Creek in central Victoria. Adapted from Ashton (1981a), by permission of the Australian Academy of Science.*

ecological drift model (Jackson 1968; Bowman and Jackson 1981; Jackson and Bowman 1982a) that stressed the importance of site characteristics. Like Jackson's model, his model was based on the assumption that fire is critical for the regeneration of all Tasmanian non-rainforest tree species. However, he suggested that vegetation patterns in western Tasmania are essentially stable and closely corresponded to site characteristics (e.g. geology, drainage, topography, climate). In his view, fire did not determine vegetation, rather the risk of fire was determined by the vegetation and was cyclical. For example, he suggested that the patterns of sedgelands and rainforest in western Tasmania are primarily controlled by edaphic factors, with rainforest occurring on the better drained soils than the sedgelands, and that fire merely reinforced such environmentally determined patterns. Unlike Jackson (1968), who stressed the role of chance variations in fire frequency, Mount believed that the occurrence of fire was determined by the amount of fuel that had accumulated since the last fire. For example, he conjectured that an unburnt patch of forest existed because 'the sparks that were blown into it in the past found insufficient fuel to start a fire – not because, by chance, sparks happen to miss it for hundreds of years'. He suggested that fires would occur periodically given appropriate climatic conditions. Further, he believed fires were effectively spread by the fall-out of sparks and burning embers entrained in updraughts above the fire front, a process he called the 'spot-fire mechanism'.

Central to Mount's fire-cycle model is the erroneous idea that fuel and fire incidence are intrinsically interdependent. Field research has not supported Mount's view that young regenerating stands of wet *Eucalyptus* forests are unlikely to burn because of low fuel levels, P. Barker (1991) supporting the prediction of Jackson's model that there is a high likelihood of fire in regenerating forest stands (i.e. a short fire-free interval). Regenerating stands have ground surface microclimates that promote rapid drying of fine fuels such as litter, fire-weeds and sapling foliage (M. Barker 1991). Bowman and Jackson (1981) argued that fuel moisture rather than fuel quantity is critical in determining the *incidence* of fire although they acknowledged that both fuel quantity and fuel moisture content influences fire *intensity*. Ellis (1965) noted that Mount's theory is seriously flawed because 'the presence of fuel alone does not ensure a fire: the location and frequency of a fire source is likely to be of even greater importance'.

Henderson and Wilkins' model led them to conclude that fires would create an ever-changing distribution of forest types in western Tasmania. Given the importance of landscape factors in controlling the distribution of fire and vegetation, their conclusion is unrealistic. They acknowledged that a realistic model would require consideration of the effect of spatial variation in fire incidence. There have been a few attempts to consider this spatial component in mathematical models (Green 1989; Bradstock *et al.* 1996; Noble and Gitay 1996). Noble and Gitay (1996) found that a two-dimensional model of the effects of fire on Tasmanian rainforest dynamics produced substantially different predictions to a model that had no spatial component. This difference appeared due to the spatial juxtaposition of fire-prone and fire-sensitive vegetation, which increased the risk of fire for the latter vegetation type. Their model ignored the effect of edaphic or topographic variation on vegetation distribution. Green (1989) found that the distribution pattern of two hypothetical vegetation types in a two-dimensional model varied in response to an environmental gradient that influenced their competitive relationships.

Random fires

The assumption that fires occur randomly is central to the ecological drift model and Henderson and Wilkin's subsequent treatment. If fires are not random, the accurate determination of fire-free intervals is severely hampered because of the peculiar mathematical properties of the Poisson distribution that describes random phenomena. If fire occurrence is strictly random then the age of unburnt stands is identical to the time interval between successive fires and the measurement of fire-free intervals involves simply determining time since the last wildfire. This can be achieved by counting the tree rings for the canopy dominants. However, if the occurrence of fire is not random, the age of fire-killed trees on burnt sites must be determined in order to measure actual fire-free intervals.

There is clear evidence that the occurrence and spatial distribution of fire is not random. Under current regimes, lightning accounts for less than 1% of the area burnt in Tasmania (Jackson and Bowman 1982b; Podger et al. 1988). Brown (1996) noted that, in southwest Tasmania, wildfires originate from people, and therefore the distribution of wildfire corresponds with areas, such as the coastline and zones of mineral exploitation, that are the focus of current economic activities. He found that river catchments in southwest Tasmania with more scrub than predicted by the Henderson and Wilkins' model were subject to European landscape burning associated with mineral exploitation. Brown and Podger (1982b) demonstrated that the occurrence of fires were 'clearly not Poisson' at Bathurst Harbour in southwest Tasmania. They found that the occurrence of fires had varied through time, with increased frequency and regularity of burning corresponding to 'protection burning' carried out by Europeans who settled in the area in the 1950s.

Given the infrequency of fires started by lightning and the prevalence of fire-induced sedgelands, it is reasonable to assume that the Holocene distribution of sedgeland in western Tasmania was largely a product of frequent, intentional Aboriginal landscape burning (Jackson 1968; Macphail 1979, 1980;

Bowman and Jackson 1981; Macphail and Colhoun 1985). There is ethnohistori-
cal and ecological evidence that Tasmanian Aboriginal people used fire to
maintain tracts of fire-prone vegetation throughout Tasmania (Jackson 1968;
Jones 1969; Macphail 1980; Macphail and Colhoun 1985; Bowman and Brown
1986; Thomas 1993, 1995). A number of authors have commented that the
periodicity and spatial pattern of fires probably changed following European
colonisation and the near extermination of the Tasmanian Aborigines (Ellis
1965; Henderson and Wilkins 1975). Clearly, the results of mathematical models
based on the assumption of random occurrence of fire cannot provide insights
into the long-term impact of purposeful Aboriginal landscape burning, although
with modification, mathematical models can be used to advance the question.
For example, Price and Bowman (1994) constructed a staged-based matrix
model to explicitly determine the combination of fire periodicities and fire
intensities required to create the demographic structures of the north Australian
conifer *Callitris intratropica* observed by Bowman and Panton (1993*a*) in the
Northern Territory. They concluded that the vast tracts of *C. intratropica* that
occur on sand sheets (i.e. sites with no topographic fire protection) in the coastal
regions of the Northern Territory are the product of skilful use of fire by
Aboriginal people.

Fire is a highly variable phenomenon

The 'ecological drift' model is built on the invalid assumption that a fire
necessarily kills all plants in any given vegetation type in western Tasmania. Hill
and Read (1984) found that a fire in a mixed *Eucalyptus*–rainforest in western
Tasmania caused the incomplete loss of adult rainforest trees and the death of
between 60% and 81% of the *Eucalyptus nitida* trees (Table 10.4). Bowman and
Jackson (1981) attempted to modify the 'ecological drift' model to take into
account the well-developed powers of vegetative regeneration of many Tasmanian plant species. A number of authors have suggested that the succession
pathways from *Eucalyptus* forest to rainforest are influenced by both fire frequency and fire intensity (Howard 1973*a*; Ashton 1981*a*; Ellis 1985; Harrington
1995). However, this approach is complicated by the requirement to accurately
characterise the intensity of a given fire. There is often huge variation in the
behaviour of even a single fire during its passage across a landscape, variation
that is related to changes in topography, fuel loads, time of day, soil types and
meteorological conditions. Gilbert (1959) described stands of *Eucalyptus regnans*
in the Florentine Valley in western Tasmania that were largely unaffected by
low-intensity fires that had burnt during the night or were quelled by rainfall.

TABLE 10.4. The pre-fire density of trees > 2 cm dbh, their post-fire survival and the density of seedlings 18 months after the fire on two transects (A and B) in 100–170-year-old mixed *Eucalyptus* forest–rainforest in the Savage River area of western Tasmania

Species	Pre-fire density (ha^{-1})		Survival of trees (%)		Density of seedlings (m^2)	
	A	B	A	B	A	B
Nothofagus cunninghamii	373	420	0	1	1.46	0.96
Eucryphia lucida	620	268	0	0	2.40	0.01
Eucalyptus nitida	124	108	42	19	0.22	1.54
Atherosperma moschatum	764	208	1	0	0.01	0
Leptospermum scoparium	1088	0	16	—	8.80	0.52
Anodopetalum biglandulosum	98	8	0	0	0	0
Anopterus glandulosus	52	68	0	0	0	0
Phyllocladus aspleniifolius	4	48	0	0	0.34	0.04
Acacia melanoxylon	0	4	—	0	0	0.72

Adapted from Hill and Read (1984).

Hill (1982) noted that the post-fire mortality of rainforest trees in western Tasmania was positively related to localised variation in peat depth, which in turn was related to the spatial distribution of individual tree species. *Atherosperma moschatum* produces leaf litter that is rapidly broken down and leads to the production of shallow organic soils. Hill (1982) found that the post-fire mortality of *A. moschatum* was less than *Nothofagus cunninghamii* and *Eucryphia lucida* because the latter produce deep organic soils that are vulnerable to peat fires. Within-fire variability renders the response of vegetation and successional processes to fire exceedingly difficult to model.

Effect of edaphic factors

Both Jackson's drift model and Henderson and Wilkins' subsequent treatment ignored the effect of soil fertility on the rate of succession following fire disturbance. There is some evidence to suggest that soil fertility can both increase and decrease the rate of succession and influence the coverage of different vegetation types in western Tasmania. Considering individual river catchments, Brown (1996) reported that infertile catchments conformed to the prediction of Henderson and Wilkins but catchments with fertile soils had greater coverage of rainforest than predicted. Read (1995) experimentally verified the effect of soil fertility on the growth rate of Tasmanian rainforest trees, and it is likely that this effect would influence the rate of succession. Bowman and Jackson (1981) modified Jackson's ecological drift model to include the effect of soil fertility on successional pathways and the rate of succession. For instance, they incorporated Kirkpatrick's (1977) field observation that long unburnt sedgelands on fertile substrates can be invaded directly by rainforest species, obviating the need for unburnt sedgelands to pass through other successional stages such as *Eucalyptus* forest. As discussed in the previous chapter, the role of soil fertility on rates of succession is complicated because fire history also affects soil fertility. In the worst case, fires can consume organic soils down to bedrock, destroying a large proportion of the soil nutrient capital (Brown and Podger 1982b; Hill 1982).

Conclusion

There are serious difficulties in realistically modelling the role of fire in controlling the distribution of rainforest in western Tasmania. Nonetheless, Jackson's (1968) ecological drift model, Henderson and Wilkins' (1975) transition probability matrix model and the 'vital attributes' models of Noble and Slatyer (1980) are all instructive because they underline the importance of fire frequency in controlling the distribution of rainforests. However, a critical assumption in these models is that fire incidence is random in space and time. This assumption is inconsistent with the idea that Aborigines may have introduced a non-random, and possible systematic and skilful, component to the occurrence of fire in the Australian landscape. Jackson (1968) was aware of the possible importance of Aboriginal burning when he wrote 'man's influence on the balance of vegetation types through his influence on the frequency of fires has had a profound effect since the Pleistocene and must continue to dominate our handling of our present resources for a considerable period'. The question we must now turn to concerns the role of Aboriginal burning in the ecology and evolution of Australian forests.

11

The fire theory IV. Aboriginal landscape burning

The realisation that fire is important in determining rainforest boundaries has naturally led many authors to speculate on the effect of Aboriginal landscape burning. This line of thinking has a long history. Domin (1911) asserted that 'secondary' *Eucalyptus* forests had replaced rainforest in southern Queensland and that this change was due to 'aboriginal inhabitants, mostly by means of bushfires'. Currently, two temporal perspectives dominate this topic. The first focuses on the possible ecological and evolutionary consequences of the arrival in Australia of a new ignition source: *Homo sapiens*. Some authors have suggested that a sharp increase in the frequency of fires caused the contraction of rainforest vegetation and possibly stimulated the evolution of fire-adapted Australian vegetation during the last ice age. For instance, Truswell (1993) concluded a review of the Cainozoic history of Australian vegetation with the thought that 'the expansion of the eucalypts, probably at the expense of the drier rainforests and Casuarinaceae woodlands, is a more recent story and has possible links with human land-management practices'. The second perspective concerns the more recent past, and argues that landscape burning by Aborigines had controlled the distribution of rainforest throughout Australia. This view is largely the product of ecological studies of rainforest boundary dynamics. Several authors consider fire as being an environmental constant through geological time (Mount 1979; Horton 1982). They see Aboriginal landscape burning as being irrelevant in determining current and past vegetation patterns.

In this chapter, I will review the evidence that Aboriginal landscape burning has had a major effect on the distribution of rainforests in the recent and distant past. I will also critically evaluate the hypothesis that their burning triggered the evolutionary diversification of non-rainforest vegetation.

Landscape ecology

Nearly all considerations of the impact of Aboriginal landscape burning on rainforests are inferential. An exception is the small study of Bowman (1992*c*) on Melville Island in the Northern Territory. He demonstrated an insignificant short-term effect of regular burning by Aborigines on isolated patches of rainforest. Nonetheless, many authors have hypothesised that a long history of Aboriginal landscape burning has determined current or recent rainforest boundaries (Domin 1911; Herbert 1938; Jackson 1965, 1968; Ellis 1985; Ash 1988; Duff and Stocker 1989; Russell-Smith *et al.* 1993; Fensham and Fairfax 1996). For example, Aboriginal landscape burning is thought responsible for the patches of treeless grasslands that occur within tracts of rainforest from Tasmania to the monsoon tropics (Jackson 1965; Ellis 1985; Ash 1988; Bowman *et al.* 1990; Harrington and Sanderson 1994; Fensham and Fairfax 1996) (Figure 11.1). The logic underpinning this inference is outlined by Ash (1988). He conjectured that 'the areal rate of fire ignition (ignitions ha^{-1} year^{-1}) required to maintain the pyrophytic vegetation boundaries is inversely proportional to the area of the patch of pyrophytic vegetation'. In other words, the smaller the patch the greater number of ignitions required for its maintenance. He noted that, given that the current incidence of lightning is insufficient to maintain pockets of fire-dependent vegetation, Aborigines must have been the source of the ignitions.

Archaeologists have debated the significance of tracts of treeless vegetation in southwest Tasmania. Cosgrove *et al.* (1990, 1994) claimed that the region was uninhabited during the Holocene because of the absence of archaeological material of Holocene age in caves that were inhabited during the Pleistocene. Thomas (1993) explained this absence of archaeological material in caves as merely a reflection of a change in local settlement patterns. He interpreted the occurrence of tracts of treeless sedgelands as a consequence of fires started by Aborigines, and thus considered the sedgelands to be reliable evidence of human occupation throughout the Holocene. Resolution of the dispute requires evidence drawn from archaeology, ecology and palaeoecology, such as the integrated study of Ellis and Thomas (1988) in the northeast of Tasmania. They concluded that Aboriginal landscape burning was responsible for grasslands pockets within large tracts of rainforest.

There is no doubt that European fire regimes are different to those of the Aboriginal period, as evidenced by the dramatic contraction of fire-sensitive rainforest trees. Landscape fires in the last 100 years have caused the loss of about 30% of the area covered by the Tasmanian endemic conifer *Athrotaxis selaginoides* (Brown 1988) (Figure 11.2). In the Australian monsoon tropics,

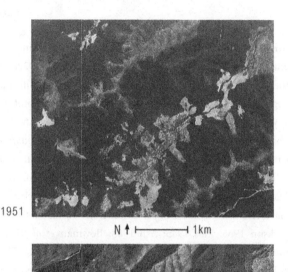

1951

N ↑ ├──────┤ 1km

1984

Figure 11.1
Aerial photographs showing the loss of grasslands, with a mosaic of rainforest (dense crown texture) and Eucalyptus *forest (light crown texture), over a 33-year period on the Bunya Mountains in southern Queensland. It is assumed that Aboriginal landscape burning maintained these grasslands. (Aerial photographs reproduced with permission of Department of Natural Resources, Queensland.)*

Allosyncarpia ternata rainforests are retreating under the current fire regime (Russell-Smith *et al.* 1993; Bowman 1994).

A number of studies have shown locally variable responses of rainforest boundaries to the cessation of Aboriginal burning. Clayton-Greene and Beard (1985) found that on islands infrequently visited by humans in the northwest Kimberley, rainforests appear to be colonising the surrounding savanna, whereas on the mainland, many rainforests have retreated in response to unfavourable fire regimes following European colonisation. They concluded that monsoon forests 'exist in a fine balance which is dependent on the fire regime, both natural (lightning) and man induced'. Podger *et al.* (1988) suggest that locally variable responses are attributable to local ecological factors such as terrain, soils and the history of land use.

In many environments, the shift from Aboriginal to European fire regimes is

Figure 11.2
Unburnt and fire-killed rainforest dominated by the Tasmanian endemic pencil pine,
Athrotaxis selaginoides. *(Photograph: David Bowman.)*

confounded by other variables. In the monsoon tropics, changes in fuels associated with introduced herbivores such as pigs, cattle, and buffalo, and the spread
of weeds, have resulted in fire damage to many rainforests (Russell-Smith and
Bowman 1992). Conversely, Harrington and Sanderson (1994) suggested that in
the humid tropics the expansion of rainforest is linked to a decline in fire
intensity associated with the change from Aboriginal fire management to pastoralism. They suggest that cattle grazing substantially reduced grass fuel loads
on rainforest margins. Graziers also intentionally imposed a regime of low
intensity fires designed to regenerate pastures. The modelling undertaken by
Price and Bowman (1994) suggested that the Aboriginal burning had maintained
stands of the fire-sensitive savanna conifer *Callitris intratropica* on sand sheets
with no topographic fire protection in the coastal regions of the Northern
Territory. Further, they suggested that a shift from low intensity to high intensity
fires associated with the cessation of Aboriginal fire management accounted for
the widespread mortality of this tree species (Bowman and Panton 1993a). There
have been no attempts to model the putative effects of Aboriginal landscape
burning on rainforest boundaries.

Historical and anthropological observations of Aboriginal landscape burning

It is remarkable how little is known about the use of fire by Aboriginal people. Historic records confirm that, at the time of European colonisation during the nineteenth century, Aboriginal people burnt landscapes for a great variety of purposes including clearance of thick vegetation to facilitate travel, signalling, controlling insects and vermin, hunting and waging war (King 1963; Hiatt 1968; Jones 1969; Stockton 1982; Kimber 1983; Hallam 1975, 1985). There is evidence of temporal and spatial variation in Aboriginal landscape fires throughout Australia (Hallam 1975, 1985; Stockton 1982; Kimber 1983; Bowman and Brown 1986; Braithwaite 1991; Fensham 1997). Fensham (1997) showed that landscape fires in Queensland were most common in winter and autumn, and that the frequency of fires was much higher in coastal and sub-coastal areas than inland areas (Figure 11.3). However, historical records can not provide accurate information on the spatial extent of fires, the types of vegetation burnt, and the reason why particular fires were lit (Hiatt 1968; Horton 1982; Bowman and Brown 1986; Fensham 1997). Perhaps the greatest limitation of historical records is our inability to determine whether Aborigines had a predictive knowledge of the ecological outcomes of burning, particularly in the longer term (Nicholson 1981; Latz and Griffin 1978; Hallam 1985; Fensham 1997). Some nineteenth century observers suggested that Aborigines used fires to achieve long-term outcomes. The explorer Major Thomas Mitchell suggested that open woodland in western New South Wales was the product of deliberate burning by Aborigines. He succinctly stated this view in an often-quoted comment that 'fire, grass, kangaroos and human inhabitants seem all dependent on each other for existence in Australia'. Mitchell argued that were it not for this burning, the landscape would be a 'thick jungle' rather than 'open forests' (Jones 1969). In aggregate, the nineteenth century record demonstrates that 'fire was the indispensable agent by which Aboriginal man extracted many of his resources from the environment' (Nicholson 1981). The ethno-historical record cannot resolve whether Aborigines had a systematic and predictive ecological knowledge concerning their use of fire, although there appears to be some evidence to support this view.

The nineteenth century anthropological literature on Aboriginal landscape burning is sparse compared to that of the nineteenth century. In 1971, the anthropologist Gould wrote that 'despite the demonstrated importance of this question, there exist as yet no holistic and empirical studies of fire as employed by any particular ethnographic hunting-and-gathering society in the world'! Whatever the cause of this anthropological neglect, the upshot for nearly all

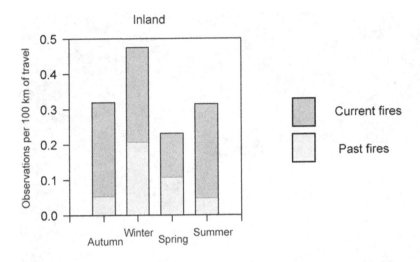

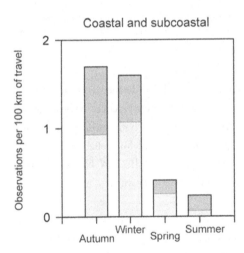

Figure 11.3
Numerical analysis of records of Aboriginal burning derived from the journals of 19th century explorers in the State of Queensland. The number of records of 'current fires' which may have been lit in response to the presence of the explorers, and records of recently burnt landscapes ('past fires') that would not have been influenced by the presence of the explorers, is shown. The records are standardised by expressing records for 100 km sections of each explorer's traverse. Geographic categories includes the following communities defined by Fensham (1997): 'Inland' – Aristida, Channel, Spinifex, Brigalow, Blue Grass; 'Coastal and subcoastal' – Blady grass, Blue grass – brown top, littoral, rainforest, Black Spear grass and Schizachyrium. Adapted from Fensham (1997).

Figure 11.4
A fire stick made from bush materials, northeast Arnhem Land, 1997. Aboriginal people in this region still possess a large amount of traditional knowledge concerning the use of fire as a land management and hunting tool. (Photograph: David Bowman.)

Australian environments is that there is a lack of 'detailed empirical studies regarding the specific effects which Aboriginal burning practices' have on vegetation (Lewis 1982). This has occurred even though both anthropologists and ecologists have noted the importance of, and need for, collaborative ethno-ecological research on Aboriginal fire usage (Gould 1971; Webb 1977; Haynes 1978; Nicholson 1981; Stocker and Mott 1981; Bowman and Brown 1986; Bowman 1995).

The published anthropological studies of Aboriginal burning in forested environments all concern the monsoonal tropics, where Aboriginal landscape burning practices have persisted longer than elsewhere (Figure 11.4). These ethnographic studies are remarkably consistent with the interpretations of the ethno-historical record discussed above (Gould 1971; Kimber 1983; Latz and Griffin 1978; Jones 1975, 1980). Thomson (1939, 1949) outlined a complex and systematic temporal variation in the use of fire, a view corroborated subsequently by Jones (1975, 1980) and Russell-Smith *et al.* (1997*a*). Jones (1975, 1980) spent one year with a group of Aborigines in their tribal lands on the Blyth River in central coastal Arnhem Land and has eloquently described systematic and skilful burning linked with the orderly seasonal exploitation of different environments.

He found that most burning occurred early in the dry season when low-intensity fires could be easily controlled using predictable wind changes and falls of nocturnal dew. The product was a complex mosaic of burnt patches that subsequently served as fire breaks, enabling the control of fires lit later in the year. Jones reported that fire-sensitive rainforest patches were protected from fire by burning fire breaks on their perimeter, and that this was done out of deference to vengeful supernatural beings believed to occupy the patches. Russell-Smith *et al.* (1997*a*) ingeniously reconstructed seasonal patterns of landscape burning in Kakadu National Park. They did this by relating the observed timing of harvests of staple food plants that grow in particular plant communities with the records of landscape burning made by 19th century European explorers. They concluded that Aborigines used fire skilfully in the course of the annual cycle of hunting and food gathering throughout their clan estates. One important consequence of such burning was the conservation of yams (e.g. *Amorphophallus paeoniifolius*, *Dioscorea* spp. and *Ipomoea* spp.) which grow on the margins of monsoon rainforests and which are an important seasonal source of carbohydrate (Figure 11.5). Russell-Smith *et al.* (1993) provided the following outline of how Aborigines may have used fires in western Arnhem Land. Burning of clan estates commenced at the end of the wet season and proceeded throughout the dry season. Curing of grass fuels and movements of family groups that tracked seasonally limited resources such as yams and game determined the timing of individual fires. Some areas were burnt to assist hunting, to clear camp sites and to make walking routes trafficable, whilst other fires were lit to create fire breaks around monsoon rainforests and other foci of valuable resources. The cumulative effect of this burning was to create a mosaic of early-burnt, late-burnt and unburnt country that prevented large conflagrations late in the dry season. Their narrative is consistent with the observation that the cessation of Aboriginal fire management has resulted in intense late dry-season fires that have caused extensive damage to monsoon rainforests (Russell-Smith and Bowman 1992).

Though the ethnographic literature is sparse, it is clear that fire was a powerful tool that Aborigines used 'systematically and purposefully over the landscape' (Russell-Smith *et al.* 1997*a*). However, such ethnographic records do not resolve the question of the long-term consequences of Aboriginal burning. Russell-Smith *et al.* (1993) noted that the long-term impact of Aboriginal burning is unknown and complicated by the great length (possibly more than 50 000 years) of human occupation of Australia. They noted that during this period there have been major changes in the density and cultural practices of Aboriginal populations, changes driven primarily by environmental change associated with the end

Figure 11.5
Harvesting yams in a recently burnt area in northeast Arnhem Land. Locating and harvesting yams and conserving their populations requires considerable skill and ecological knowledge, and remains an important activity for Aboriginal people throughout northern Australia. (Photograph: David Bowman.)

of the ice age. For insights into the question of long-term effects of Aboriginal burning, we must consider the available palaeoecological evidence.

Palaeoecology

Assessment of palaeoecological evidence of an effect of Aboriginal landscape burning on the distribution of Australian rainforests requires consideration of the following three time periods:

(i)   The last glacial cycle that commenced some 100 000 years ago. The late Pleistocene had colder climates that were initially more humid and subsequently more arid than the present. Humans colonised Australia some time during the middle of this period and before the onset of the severe aridity.

(ii) The end of the last ice age some 10 000 years ago. The

Pleistocene-Holocene boundary provides insights into the combined effects on rainforests of climatic amelioration and Aboriginal landscape burning.

(iii) The current interglacial from about 10 000 years ago to the present (the Holocene). Sea levels, climates and Aboriginal populations changed and eventually stabilised towards the end of this period. However, over the last 200 year period there has been an abrupt transition from Aboriginal to European land management.

The following discussion commences with the recent past and then works backwards into the more distant, less knowable, past.

The Holocene

Pollen evidence demonstrates a change in the fire regime following the colonisation of Australia by Europeans, although the direction and magnitude of this change varies between environments (Head 1989). For example, Gell *et al.* (1993) have shown that the arrival of Europeans in the Delegate River catchment in southeastern Victoria caused a massive increase in microscope charcoal in swamp sediments. Not until fire control measures were instituted in the 1940s did microscopic charcoal decrease. On the other hand, Dodson *et al.* (1993) demonstrated a decrease in microscopic charcoal following European colonisation at two sites on the southeast coast of New South Wales. This variable response is consistent with the landscape ecology evidence discussed previously. A number of authors have noted that caution is required in attributing vegetation changes to the transition from Aboriginal to European fire regimes because of complex environmental changes associated with European settlement, including the introduction of alien plants and animals, land clearance and subsequent soil erosion (Horton 1982; Head 1989; Boon and Dodson 1992; Dodson *et al.* 1993; Pulsford *et al.* 1993; Hope 1999). For example, Russell-Smith (1985*b*) found palaeoecological evidence that rainforests on the margin of floodplains in Kakadu National Park had contracted in the last 100 years. He attributed these changes to the combined effect of increased severity of fires and the destructive impact of feral water buffalo (*Bubalus bubalis*).

Prior to European colonisation, the effect of Aboriginal burning on Australian vegetation was not uniform (Head 1989). Dodson *et al.* (1986) concluded that fire did not become an important feature of the environment at Barrington Tops on the central coast of New South Wales until about 3000 years BP, at which time there was a decline in *Nothofagus moorei* rainforest and possibly, *Casuarina* in

Eucalyptus forests. However, they related the change to a change in climate rather than a change in burning by Aborigines because of the lack of archaeological sites attributable to this time. Over the last 2000 years, the climate has apparently been relatively stable and there has been relative stability in the ratio of *Nothofagus to Eucalyptus* pollen (Figure 11.6). Dodson *et al.* (1994) concluded that Aboriginal burning had little impact on the distribution of rainforest and tall *Eucalyptus* forest at Barrington Tops during this time. Considering Tasmania as a whole, Macphail (1991) interpreted pollen evidence as showing that climate change and Aboriginal burning caused substantial shifts in the distribution and abundance of rainforest taxa. The mid-Holocene decline of *Nothofagus cunninghamii* in central but not western Tasmania was apparently due to a steepening of the rainfall gradient from the humid west coast to the drier east coast of the island. He suggested that the increased frequency of fires during the mid Holocene limited the ranges of rainforest tree species such as *Athrotaxis, Lagarostrobos franklinii, Nothofagus gunnii* and *N. cunninghamii* (Macphail 1980, 1984, 1991). However, charcoal particle evidence from a pollen core where *Athrotaxis* and *Diselma* become locally extinct is not convincingly linked to increased charcoal concentrations, contrary to Macphail's (1991) interpretation (Figure 11.7).

Little is known about changes in the distribution of monsoon rainforests during the Holocene because of the dearth of sites suitable for the preservation of palaeoecological materials. One exception is the study of Shulmeister (1992) of late-Holocene vegetation changes in an area surrounding a coastal swamp on Groote Eylandt. He suggested that an increase in the density of Aboriginal population may have caused some vegetation changes, but this effect must have been minor given that rainforest patches were a component of the contemporary landscape he studied. An alternative means of exploring the dynamics of monsoon rainforest boundaries has been by dating abandoned Orange-footed Scrubfowl mounds. Scrubfowl are large-footed ground-dwelling rainforest birds which incubate their eggs using heat from soil and the decomposition of leaf litter raked up into 'nests' sometimes more than one metre high (Crome and Brown 1979). Scrubfowl cannot construct mounds in savannas (Bowman *et al.* 1994*b*). Stocker (1971) noted a positive correlation between distance from rainforest patches and radiocarbon ages of abandoned Orange-footed Scrubfowl mounds from one locality on Melville Island. He hypothesised that the rainforests had gradually retreated during the Holocene in response to climate change, cyclonic storms, Aboriginal landscape burning or a combination of these variables. Given the patchy geographic distribution of abandoned Scrubfowl mounds, it is unlikely that climate change caused the contraction of the monsoon rainforests (Stocker

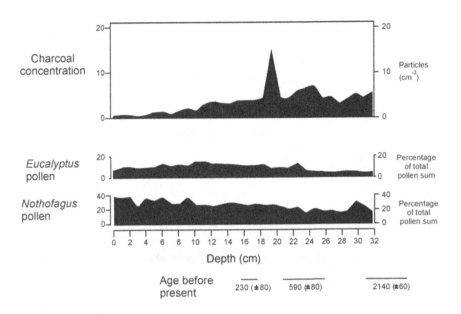

Figure 11.6
Variation during the last 2000 years in the abundance of Nothofagus *and* Eucalyptus *pollen and the density of microscopic charcoal in sedimentary samples from a core from Burraga Swamp on the Barrington Tops on the central coast of NSW. Adapted from Dodson et al. (1994).*

1971; Bowman *et al.* 1994*b*). Monsoon rainforest patches have variously contracted or expanded under contemporary climates, and there is no geomorphological evidence for substantial climatic change in northern Australia during the late Holocene (Woodroffe *et al.* 1985, 1993; Woodroffe and Chappell 1993). It is also unlikely that Aboriginal burning caused the regional contraction of monsoon rainforest because rainforest boundaries have been maintained under a regime of Aboriginal burning (Jones 1980; Bowman 1992*c*). Bowman *et al.* (1999*a*) argued that the cause of localised contraction of monsoon rainforest is the combined effect of tropical cyclone damage and subsequent severe fires in the storm debris. This suggestion does not preclude a role for Aboriginal people. It is possible that heavy fuel loads created by cyclonic storms exceeded the capacity of traditional Aboriginal fire management to avoid destructive fires. It is possible that Aborigines intentionally set fire to cyclone debris but this is inconsistent with the limited ethnographic evidence that suggests they conserved fire-sensitive vegetation.

In summary, there appears to be no clear-cut evidence that Aboriginal burning caused a significant contraction of Australian rainforests during the Holocene.

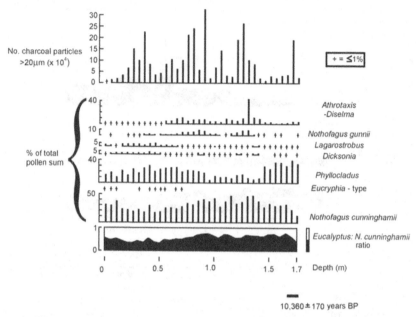

Figure 11.7
Variation during the Holocene in the abundance of selected Tasmanian rainforest trees, the ratio of Eucalyptus *and* Nothofagus *pollen and the density of microscopic charcoal in sedimentary samples in a core from the Denison Range in southwest Tasmania. Adapted from Macphail (1991), Commonwealth of Australia copyright reproduced by permission.*

The Pleistocene–Holocene boundary

There is good palynological evidence that rainforest expanded throughout eastern Australia at the end of the Pleistocene (Macphail 1975; Dodson *et al.* 1986; Walker and Chen 1987). Macphail (1991) documented a Tasmania-wide expansion of rainforest in the early Holocene in response to climatic amelioration (Figure 11.8). Climatic warming at the beginning of the Holocene resulted in rainforest establishing in the eastern Barrington Tops about 7000 years BP (Dodson *et al.* 1986). Tropical rainforest on the Atherton Tablelands is thought to have expanded at the end of the Pleistocene due to increased rainfall. However, pollen data from six sites in close proximity revealed that the timing of expansion varied by up to 3000 years (10000 to 7000 years BP) (Walker and Chen 1987). Once initiated, however, the transition to rainforest occurred relatively rapidly (<1500 years). A number of dominant taxa showed exponential population growth through this transition period, with doubling times for populations between 63 and 165 years for angiosperms and 199 and 355 years for gymnosperms (Chen 1988). Fine-resolution palynology from a core from Lake Barrine

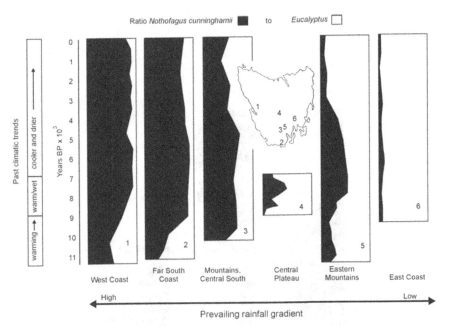

Figure 11.8
Variation during the Holocene in the ratio of Nothofagus cunninghamii *and* Eucalyptus *pollen in sedimentary samples from six selected Tasmania pollen cores. The direction of the prevailing rainfall gradient and the trends in the Holocene climates of Tasmania are also indicated. Adapted from Macphail (1991), Commonwealth of Australia copyright reproduced by permission.*

showed that the expansion of rainforest in the first half of the Holocene was associated with a decrease in microscopic charcoal particles (Walker and Chen 1987) (Figure 11.9). Under the late Pleistocene (> 9300 years BP) climate, which did not favour the establishment of rainforest, the periodicity of fires was estimated to be about one per decade. The frequency of fires decreased to about one every 220–240 years in the early Holocene (9300 to 6800 years BP), when a moister climate favoured rainforest establishment. When rainforest had taken over by the end of the early Holocene, the fire frequency declined dramatically. Walker and Chen suggested that rare deviations from the mean fire-free interval may have allowed the rainforest to establish, and that once established, the risk of fire may have decreased sharply. They noted that the cause of these departures from the mean fire frequency is unknown, but that the causal factors must have been site specific given the regional climatic conditions had been suitable for rainforest establishment for several thousand years. One possible source of local variation in fire regimes they considered was Aboriginal landscape burning.

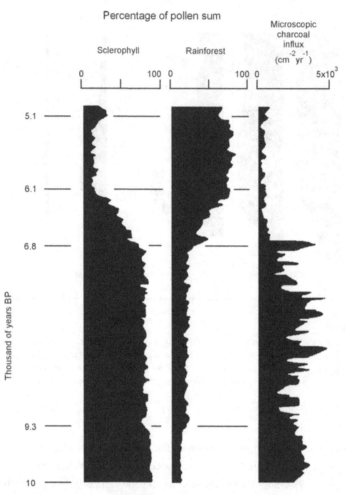

Figure 11.9
Variation during the late Pleistocene–early Holocene boundary in the abundance of rainforest and non-rainforest pollen taxa and the density of microscopic charcoal particles during the Holocene in a core taken from Lake Barrine on the Atherton Tablelands in northeast Queensland. Adapted from Walker and Chen (1987), with permission from Elsevier Science.

Macroscopic charcoal evidence provides support for the work of Walker and Chen (1987). Hopkins *et al.* (1993) suggested that climatic amelioration in the early Holocene was primarily responsible for the expansion of rainforest on the Atherton Tablelands. However, they acknowledged that rainforest expansion was staggered, or 'time-transgressive', reflecting the importance of local site factors such as rainfall, topography, soils, and proximity to adjoining *Eucalyptus* forests and thus exposure to high fire risks. They noted that the cause of the fires is unknown, but that fires lit by Aborigines may have influenced the rate of rainforest colonisation during the Holocene. They pointed out that, given that lightning is the cause of some forest fires under current climatic conditions, it might also have been important in the early Holocene.

The widespread expansion of rainforest in the adjacent tropical lowlands took place more recently, during the late Holocene (Hopkins *et al.* 1996). The concentration of numerous rare, monotypic, and endemic plant species clearly demonstrates that Holocene fires did not cause the total destruction of lowland rainforest (Webb and Tracey 1981). Hopkins *et al.* speculated that the delay in the establishment of lowland rainforest may have been a consequence of an increase in burning associated with an increase in the density of Aboriginal populations in the lowlands during the Holocene, perhaps triggered by the drowning of the continental shelf during the early Holocene. Aborigines may have used fire to create habitat mosaics in tracts of rainforest to favour game species such as the red-legged pademelon (*Thylogale stigmatica*), a macropod of rainforest ecotones (Vernes *et al.* 1995). Such burning may have caused some retreat of rainforest under late-Holocene climates that Hopkins *et al.* considered drier and less favourable for rainforests than the climates of the early and mid Holocene. It is important to note that rainforests provided valuable resources for some Aboriginal tribes, providing an incentive for the preservation of this habitat. Archaeological research has demonstrated that by 5000 years BP and possibly much earlier, some Aboriginal tribes had adapted to life in rainforest environments (Cosgrove 1996).

Although there is evidence that Aboriginal burning influenced the establishment of rainforest at the end of the last ice age, it is clear that Aboriginal burning was unable to reverse the expansion of rainforest.

The late Pleistocene

Three long pollen cores have been interpreted as demonstrating that Aboriginal burning caused the contraction of rainforests in eastern Australia in the late Pleistocene. They are: (a) Lynchs Crater on the Atherton Tablelands in the humid tropics of northeastern Queensland; (b) a deep sea core (ODP 820) about 80 km to the northeast of Cairns also in far north Queensland; and (c) Darwin Crater in southwest Tasmania. In all cases, the decline of rainforest pollen and the corresponding increase of pollen derived from fire-adapted species and microscopic charcoal has been attributed to landscape burning by colonising Aborigines. The evidence from each core is discussed below.

Lynchs Crater

Kershaw (1974, 1976, 1978, 1985, 1986) has interpreted the vegetation history from a core taken from Lynchs Crater as registering the impact of landscape burning by colonising Aborigines. The core is thought to contain sediments accumulated over the last 190 000 years, spanning two glacial–interglacial cycles (Figure 11.10). Kershaw demonstrated that rainforest of the mesophyll vine forest type (MVF), which is currently dominant on the Atherton Tablelands, occurred only twice in the pollen record (Figure 11.10, zones G2 and E3). The occurrence of rainforest of the notophyll vine forest type (NVF, zones G1 and E2), or a combination of MVF and NVF, was interpreted by Kershaw (1986) as reflecting slightly drier climates than today. During most of the period covered by the pollen record, the vegetation was thought to be notophyll/microphyll vine forest (zones D3, D1, D2, C and F) dominated by *Araucaria*, a southern hemisphere rainforest conifer with sclerophyll foliage. Commencing at around 38 000 years BP, the vegetation changed from *Araucaria*-dominated rainforest to *Eucalyptus* (Figure 11.11). This transition was accompanied by a dramatic increase in microscopic charcoal, a decline in the rainforest tree genus *Podocarpus*, and the apparent extinction of the rainforest conifer *Dacrydium*. Kershaw (1986) argued that this vegetation change was the result of Aboriginal burning, discounting the role of climate change for the following reasons. Firstly, he believed that 'a change in climate alone could not have brought about this [vegetation] change because of the potential of 'drier' rainforest to survive mean annual precipitation levels as low as 600 mm on basaltic soil such as those surrounding the site, and because of the extinction of *Dacrydium* and perhaps other taxa which must have been equipped to survive the vicissitudes of Quaternary climates'. Secondly, the vegetation transition appeared to have occurred gradually over some 12 000 years (zone C). Thirdly, a *Eucalyptus*-dominated pollen assemblage had not been previously recorded in the core. Indeed, only a few grains of pollen attributable to *Eucalyptus* had been recorded in the late Tertiary–early Pleistocene pollen record for the Atherton Tablelands (Kershaw and Sluiter 1982). Pollen data from Strenekoffs Crater on the Atherton Tablelands have been interpreted as demonstrating the impact of Aboriginal burning on late Pleistocene rainforests, although its discontinuous stratigraphy prohibits a reliable chronology (Kershaw *et al.* 1991*b*; Kershaw 1994*b*).

Kershaw's interpretation that fire caused the contraction of *Araucaria* rainforest appears to be consistent with the available ecological data. Abrupt *Araucaria cunninghamii* and *A. bidwillii* boundaries with grassland and *Eucalyptus* forest have been attributed to the impact of wildfire (Cromer and Pryor 1943; Webb

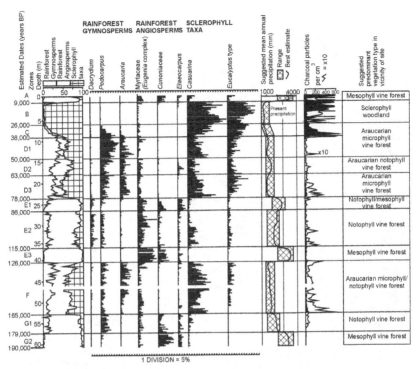

Figure 11.10

Variation in the abundance of selected pollen taxa and density of microscopic charcoal (exaggerated by a factor of 10) during the last two glacial–interglacial cycles in samples taken from a core through sediments in Lynchs Crater on the Atherton Tablelands in far north Queensland. The diagram is divided up into zones (A, B, C, D1, etc.) that are thought to reflect distinct pollen assemblages. For each of these zones, the inferred annual rainfall, vegetation type (following Webb's [1959] physiognomic scheme) and estimated chronology is also shown. Reprinted with permission from Nature *(Kershaw 1986). Copyright (1986) Macmillan Magazines Limited.*

1964; Fensham and Fairfax 1996). Although Clark (1983) has queried the ecological significance of the Lynchs Crater microscopic charcoal data, Kershaw's interpretation of these data has been subsequently supported by radiocarbon dating of macroscopic charcoal from rainforest soil in the highlands of northeast Queensland. Macroscopic charcoal studies have shown that *Eucalyptus* forests, and associated fires, were widespread during the late Pleistocene (Hopkins *et al.* 1990, 1993). Hopkins *et al.* (1993) suggested that the oldest dates reflect *Eucalyptus* forest expansion into rainforest, whilst the youngest dates correspond with the re-expansion of rainforest (Figure 11.12). As noted in the previous section, Hopkins *et al.* (1993) suggested that the variation in the timing of rainforest expansion and contraction reflected the importance of local site

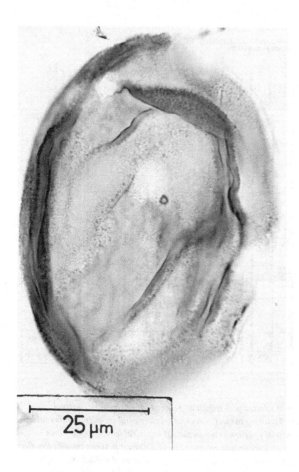

Figure 11.11
A fossil Araucaria *pollen grain. Such microscopic particles are the basis for most historical interpretation of the dynamics of Australian rainforest. (Photograph: Michael Macphail.)*

factors. Sites that currently receive low rainfall had greater variation in the age of *Eucalyptus* charcoal than sites that currently receive high rainfall. They interpreted this pattern as showing that humid sites were invaded last by *Eucalyptus* and were the first sites to be re-colonised by rainforest. Kershaw believed that Aborigines were the cause of the decline of rainforest in the late Pleistocene. However, Hopkins *et al.* acknowledged that the expansion of *Eucalyptus* forest in the late Pleistocene may have been linked to drier and possibly more frost-prone climates that favoured fire. They note that both Aborigines and lightning may have been important ignition sources. It is impossible to determine cause and effect from the charcoal record. Increased burning may have *caused* the decline of rainforest or increased burning and the establishment of flammable non-rainforest vegetation may be a *consequence* of climate change (Ladd 1988).

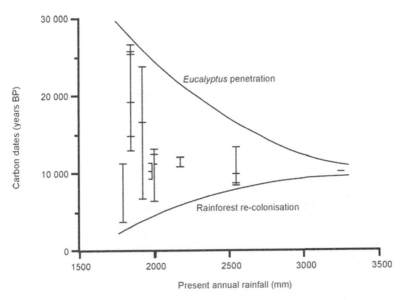

Figure 11.12
Relationship between the variation in age of macroscopic charcoal found in rainforest soil as estimated by radiocarbon dating (short horizontal lines) and present annual precipitation for eight sites in the highlands of northeast Queensland. Vertical lines join radiocarbon dates from the same site. Adapted from Hopkins et al. (1993).

ODP 820

Kershaw *et al.* (1993) examined a continuous pollen record from a deep-sea core (ODP Site 820) 80 km from the northeast coast of Queensland and thought to have developed following 1.5 million years of sedimentation (Figure 11.13). They suggested that the waxing and waning of *Araucaria*-dominated rainforest, tropical rainforest and freshwater swamp vegetation corresponds to global fluctuations in climate and sea level, but prior to the beginning of the last glacial there was no evidence of *Eucalyptus*. At this time, *Araucaria* rainforest was replaced by taxa thought indicative of *Eucalyptus* forests. Abundant microscopic charcoal particles and the disappearance of *Dacrydium* pollen accompany this transition. This floristic transition is in broad agreement with the trends determined for Lynchs Crater. A key exception is that the abundance of *Podocarpus* pollen in the deep-sea core was not correlated with charcoal particle abundance. Although Kershaw *et al.* noted the possibility that these changes could reflect a long-term trend of 'increasingly drier and more variable climatic conditions' operative since the mid Tertiary, they favoured the hypothesis that the transition was a response to Aboriginal landscape burning. They were particularly struck by the

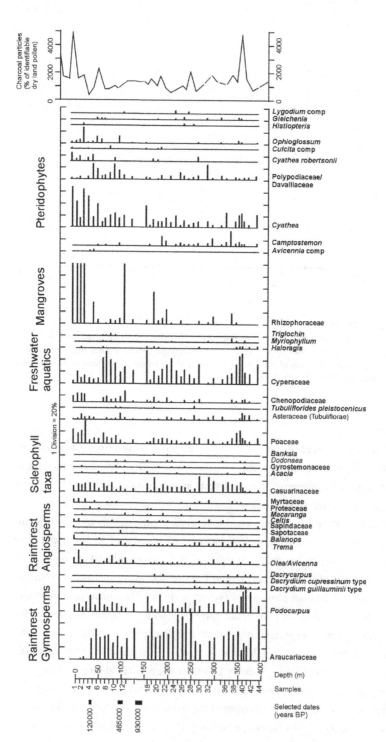

Figure 11.13

Variation during the Quaternary in the abundance of pollen taxa and density of microscopic particles from the deep sea core 820 to the northeast of Cairns. Three estimated dates are shown. Adapted from Kershaw et al. (1993).

coincidence of the vegetation change in ODP 820 with those described for the Lake George pollen core by Singh and Geissler (1985).

The pollen record from Lake George in southeastern Australia is thought to span the last 350 000 years. Unlike the previous interglacials, where *Casuarina* was the dominant woody plant, the vegetation that developed about 128 000 years BP was dominated by *Eucalyptus* and associated with a peak in microscopic charcoal. Singh and Geissler (1985) interpreted this change as a response to the commencement of Aboriginal burning. However, a number of authors have disputed this (Horton 1982; Clark 1983; Ladd 1988; Head 1989).

Anderson (1994), Hope (1994*a*) and White (1994) argued that the pollen and charcoal data from ODP 820 cannot be used to bolster the theory that colonising Aborigines destroyed large tracts of rainforest with fire in the late Pleistocene. Critics have argued that the pollen evidence can be interpreted without invoking human causation, and that the data are seriously compromised by sampling and taphonomic problems. For example, White (1994) noted that the marine sediments might have been deposited intermittently so that variation in pollen or charcoal concentrations may be artefactual. Kershaw (1994*a*) rejected this criticism and argued that the core provides a reliable regional signal of vegetation change.

Darwin Crater

The long pollen core from Darwin Crater in southwest Tasmania is thought to span five glacial–interglacial cycles (Colhoun and van de Geer 1988). Throughout this core, the occurrence of *Eucalyptus* pollen is positively correlated with peaks in microscopic charcoal. Significantly, at the period of known Aboriginal colonisation of Tasmania around 35 000 years BP, there is no evidence of a massive increase in burning or of a major contraction of rainforest. This is in contrast to the interpretation of Lynchs Crater, ODP 820 and Lake George. Jackson (1999) has speculated that the increase in charcoal and decrease in the pollen of rainforest taxa in the penultimate glacial (some 160 000 to 180 000 years BP) is evidence of burning by colonising Aborigines. He supports this extraordinary claim by noting that the primacy of fire in Tasmanian ecology demands the Aboriginal colonisation and attendant increased occurrence of fire must have occurred before the last glacial. If colonisation occurred during the last glacial then there would not be sufficient time for the flora to have adapted to a high frequency of burning by Aborigines. Colonisation of Tasmania by Aborigines during the second last glacial is not supported by any other evidence of human colonisation. Indeed, if Jackson's theory were correct it would require a total revision of orthodox models of human evolution and palaeogeography. In this context, Anderson's (1994) comments about the ODP 820 core are particularly pertinent. Anderson noted that 'where it is asserted that palynological data documents pre-archaeological evidence of occupation the argument loses plausibility in proportion to its scale; extensive proxy suggestions of human activity, without any direct evidence, indicates rather the operation of natural agencies'.

General discussion

It is reasonable to conclude that discussions about the initial impact of Aboriginal landscape burning on the Australian biota are 'ahead of their time' (Bowman 1998). This is because there is so much uncertainty about the timing of Aboriginal colonisation of Australia and the timing of their alleged ecological impacts. The most conservative date for Aboriginal colonisation is around 40 000 year BP, the upper threshold of radiocarbon dating (Allen 1994; Allen and Holdaway 1995). However, Chappell et al. (1996) argued that the alternative techniques of luminescence dating should be relied upon because it can date much older sediments. Dates derived from thermoluminescence and optical luminescence extend the period of occupancy by humans of Kakadu National Park in northern Australia back to 50–60 000 year BP (Roberts et al. 1990, 1994). However, this date is not universally accepted (Hiscock 1990; Bowdler 1991; Allen 1994; Allen and Holdaway 1995; O'Connell and Allen 1998). Thermoluminescence dating has suggested that humans were present in northwestern Australia as far back as about 120 000 years BP or more (Fullagar et al. 1996), although recent work has cast doubt on the reality of this finding (O'Connell and Allen 1998; Spooner 1998). The timing of settlement across the continent is also poorly known. Research has shown that southwestern Western Australia was occupied by about 40 000 years BP (Pearce and Barbetti 1981), Tasmania by about 35 000 years BP (Cosgrove 1989), and central Australia by 27 000 years BP (Smith 1987).

There is also great uncertainty about the timing of vegetation changes attributed to landscape burning. The chronologies of the pollen cores beyond the range of radiocarbon dating have to be estimated. Singh and Geissler (1985) and Kershaw (1986) used the following method: (i) each given pollen assemblage was related to an extant vegetation type; (ii) this extant vegetation type was used as a proxy for a generalised regional climate; (iii) the regional climate estimates are matched to temperature signals interpreted from oxygen-isotope records preserved in marine cores (Shackleton and Opdyke 1973); and (iv) the chronology of the deep sea core is used to estimate the age of the pollen assemblage. Such long chains of reasoning are vulnerable to serious error given the difficulties in matching pollen assemblages to specific vegetation and specific vegetation types to regional climate estimates.

There are a number of reasons why a pollen assemblage is not a perfect record of the surrounding vegetation (Kershaw 1986; Martin 1987; Walker and Chen 1987; Macphail *et al.* 1993). They include the following: (i) wind-dispersed pollen has a greater likelihood of preservation than animal-dispersed pollen; (ii) the effectiveness of pollen capture may change in response to changes in the sedimentary body (e.g. change from a lake to a swamp); (iii) small pockets of vegetation such as rainforest in a gorge, may be invisible palynologically; (iv) the relationship between the abundance of a given pollen type and the abundance (biomass, density etc.) of source taxa is highly variable; (v) identical pollen morphologies may be derived from taxa with different ecologies; and (vi) the ecological relationships of a group of species with identical pollen morphologies may have changed through time (indeed species driven to extinction may have pollen that is the same as extant species). Some of these problems can be illustrated from the Lynchs Crater core (Figure 11.10). During interglacial periods (Figure 11.10, zones G2, E3 and A), there were substantial fluctuations in the abundance of Myrtaceae pollen in the *Eugenia* complex. Kershaw (1986) accounted for these fluctuations as 'at least partly related to the inclusion in the taxon of several different species with various ecological requirements and differential pollen production and dispersal capabilities'. He explained the increased ratio of gymnosperm to angiosperm pollens through the late Pleistocene as being a consequence of the filling of Lynchs Crater. It is thought that Lynchs Crater was originally a deep lake that was effective in capturing local sources of pollen, but subsequent infilling produced a swamp environment that favoured the capture of pollen derived from regional sources.

Given the above problems, it should come as no surprise that the age of charcoal peaks outside the range of radiocarbon dating have been disputed. Consider the Lake George core. Wright (1986) used a simple linear regression of depth with radiocarbon dates to estimate that the age of the transition from *Casuarina* to *Eucalyptus* is 54 000 years BP rather than 128 000 years BP as suggested by Singh and Geissler (1985). However, Wright's (1986) interpretation may be flawed because it is based on the assumption of a constant rate of sedimentation (Kershaw *et al.* 1991*a*; Kershaw 1994*a*). It is not necessarily true that there is a linear relationship between sample depth and sample age because sedimentation rates may change in response to the geomorphological evolution of the site containing the pollen core.

These uncertain chronologies make it impossible to know if the lack of synchronicity between estimated ages for microscopic charcoal peaks from the ODP 820, Lynchs Crater, Darwin Crater and Lake George cores is real or artefactual. If the asynchrony is real, this is problematic for the hypothesis that Aboriginal colonisation in the Pleistocene was characterised by the large-scale destruction of rainforests with fire. A number of complicated explanations of the asynchronicity of alleged Aboriginal burning impacts have been advanced (Singh *et al.* 1981; Kershaw 1981, 1984, 1994*a,b*). For example, Singh *et al.* (1981) explained the different age of charcoal peaks in Lynchs Crater and Lake George in the following way. They wrote that 'the last interglacial in north-eastern Queensland has supported a dense rainforest which man may have found difficult to penetrate until the change to a significantly drier climate during the last glacial period. On the other hand, in the temperate south man would have found the warm interglacial climate and the developing sclerophyll vegetation quite conducive to his brush-firing activities'. Kershaw (1994*a*) argued that the asynchrony between ODP 820 and both Lynchs and Strenekoffs Craters was the result of Aboriginal burning that commenced on the coast and proceeded gradually to the hinterland rainforests. He suggested that the rate of conversion from rainforest to *Eucalyptus* forest might have been slower in the highlands than the lowlands because of the moister climate.

A basic problem with the hypothesis that colonising Aborigines systematically destroyed massive areas of rainforest with fire concerns the lack of motive. It has been suggested that the first colonists arrived from southeast Asia culturally adapted to life in rainforests (Bowdler 1983). Equally, it is not far-fetched to assume that the first Aboriginal colonists of Australia were already skilled in the use of fire (Jones 1979; Head 1989). It has been suggested that anthropogenic burning produced habitats that facilitated the spread of cattle (*Bos* species) into southeast Asian regions during the Pleistocene (Wharton 1969). The first people may have used fire to create habitat mosaics to increase the abundance of game and food plants. Even assuming a lack of appropriate knowledge, Head (1989) suggested that Pleistocene Australians might have minimised the impact of their fires once they became familiar with the Australian environment.

Although the oldest archaeological material found in Australia has been in the monsoonal tropics, little is known about the initial effect of the colonists on the environment; indeed almost nothing is known about the vegetation history of this region during the Pleistocene. Although extrapolation of historic biogeographic trends from humid to monsoonal environments has sometimes been attempted (Kershaw 1985), it is difficult to justify. The ecological and floristic complexity of the monsoon savanna vegetation suggests that it is ancient and that the first people colonised landscapes dominated by savanna rather than rainforest. Whatever the dominant habitat, the strong similarities amongst floras of the Old world tropics including northern Australia means that first colonists would have been familiar with most of the northern Australia plant genera as well as some of the plant species (Golson 1971; Bowman *et al.* 1988*a*; Liddle *et al.* 1994).

Aboriginal landscape burning as an agent of evolution

A number of authors have hypothesised that a long history of burning by Aborigines not only caused the geographic expansion of non-rainforest vegetation but also triggered the evolutionary diversification of fire-adapted species (Kershaw 1981, 1984; Singh *et al.* 1981; Gillison 1983; Janzen 1988; Beard 1990; Hill 1994*c*) (Figure 11.14). Kershaw (1981) argued that 'the recent increased impact of fire is probably related to an increase in importance of fire-promoting species – particularly the eucalypts – within the flora. Although it is very unlikely that eucalypts evolved as recently as this, it is possible that the ecology of the genus has changed through time. The great variation shown within species and species groups would support a relatively recent radiation and massive extension of range which may have been linked with the development of open eucalypt communities as we see them today'. Kershaw concluded that 'the most tenable explanation of why [these] changes occurred at this particular time is that aboriginal man encouraged fire and its effects'.

This evolutionary hypothesis postulates a quite rapid ($< 10^5$ years) diversification of the fire-adapted biota. The alternative hypothesis is that the diversification of the non-rainforest biota occurred over a much longer period (10^6 or 10^7 years) and was largely independent of Aboriginal use of fire. The fossil record is unable to provide detailed insights into this problem because the moisture-limited environments have a poor fossil record that 'precludes the full elucidation of the origin, nature, and distribution of sclerophyll and drier rainforest vegetation' (Kershaw 1984). Evidence from ecological biogeography, however, strongly suggests that the distribution and diversity of the Australian non-rainforest vegetation developed over a very long period. The evolution of *Eucalyptus* substantiates the point.

Evolution of *Eucalyptus*

There are probably a little less than a thousand species of *Eucalyptus*, most of them confined to Australia but with a few species to the north in Timor, New Guinea, Sulawesi and Mindanao (Pryor and Johnson 1981; Ladiges 1997). Here I

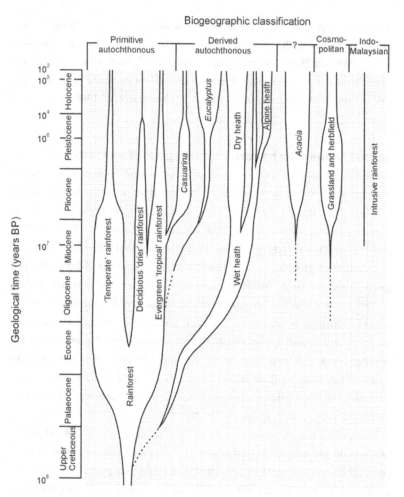

Figure 11.14
A diagrammatic representation of the assumed evolutionary relationships and patterns of diversification of dominant plant taxa and major Australian vegetation types. Adapted from Kershaw et al. (1991b).

use *Eucalyptus* in the broad sense, including species recently attributed to the genus *Corymbia*.

The gross morphological similarity of *Eucalyptus* foliage, flowers, fruits and bark types does not faithfully reflect the enormous anatomical, developmental and biochemical variation within the genus. Following a review of phylogenetic relationships as inferred by morphology and DNA, biogeographic patterns and fossil evidence, Ladiges (1997) concluded that *Eucalyptus* and related genera form a taxonomic complex reflecting a long evolutionary history. The prevalence of hybrid swarms and clines has been cited as evidence that the genus is very recent and still rapidly diversifying (Barber and Jackson 1957; Potts and Reid 1990). Ladiges argued instead that the role of hybridisation in forming new species is equivocal, and that diversification within the genus has been primarily by vicariant speciation. The meagre fossil record of *Eucalyptus* demonstrates that the genus pre-dates the separation of New Zealand and New Caledonia from the Australian plate some 60 million years ago. *Eucalyptus*-like fruits and foliage macrofossils of mid-Tertiary age have been found in New Zealand and in many Australian localities (Pole 1993; Rozefelds 1996). Recently, 'immaculately preserved mummified *Eucalyptus* fruits' have been discovered in late-Miocene deposits in Victoria (Rozefelds 1996).

A feature of eucalypt communities is their considerable structural and floristic diversity (Wardell-Johnston *et al.* 1997). *Eucalyptus* dominates all the forested biomes in Australia that form a discontinuous arc from southwest Western Australia across the eastern seaboard to the Kimberley region in northwestern Australia. *Eucalyptus* communities also dominate montane environments excepting sites subjected to extremely low temperatures. Eucalypts form stunted woodlands called 'mallee' in some semi-arid environments. Throughout arid central Australia, the genus is ecologically subordinate to *Acacia* and the endemic Australian grass *Triodia* (Bowman *et al.* 1994a; Bowman and Connors 1996). The growth habit of *Eucalyptus* is highly variable and corresponds with habitat types. The smallest *Eucalyptus* is an endemic Tasmanian alpine shrub (*Eucalyptus vernicosa*) that reaches maturity when less than 0.5 m in height (Williams and Potts 1996). The tallest eucalypt species, *E. regnans*, can reach heights of more than 100 m on fertile and perennially moist soils (Williams and Potts 1996), qualifying this species as the tallest of all extant angiosperms. *Eucalyptus* in semi-arid or montane environments often have massive lignotubers that give rise to numerous short stems rather than a single trunk, a habit termed 'mallee'. Eucalypt species have clear environmental specificity (Austin *et al.* 1997). For instance, Hughes *et al.* (1996) found that 67% of Australian *Eucalyptus* species have geographic distributions for which the range of mean annual temperatures is less than 5°C, and that 25% of these species have a thermal range of less than 1°C (Figure 11.15). *Eucalyptus* species often show strong correlation with specific edaphic conditions. For example, of 21 common species of *Eucalyptus* recorded in the upper reaches of the South Alligator River in the Northern Territory only four of these species did not show statistically significant differences in basal area between distinct edaphic environments (Bowman *et al.* 1993). At sites where there is a high diversity of *Eucalyptus* species there is also evidence of clear

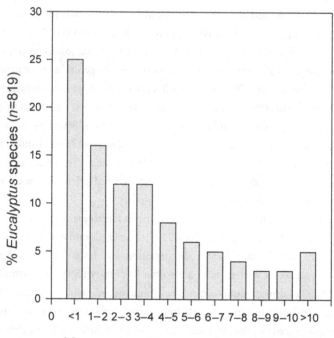

Figure 11.15

Proportion of Eucalyptus *species (n = 819) in 12 categories of geographic thermal ranges as determined by the range (in °C) of the estimated mean annual temperature of all sites where a given species has been recorded. Adapted from Hughes et al. (1996).*

separation in flowering times among those species (Bowman *et al.* 1991*b*). All these facts are consistent with a long evolutionary history.

The distribution patterns of eucalypts are also consistent with the view that *Eucalyptus* is an ancient Australian genus. Consider *Eucalyptus tetrodonta*. It occurs in a series of large isolated tracts from Cape York Peninsula in Queensland to Cape Leveque in Western Australia, a distance of over 2000 km (Chippendale and Wolf 1981). It is improbable that this distributional pattern is a consequence of long-distance dispersal because this species has a poorly dispersed small spherical seed of about 1 mg in weight. Furthermore, under contemporary environmental conditions, seedling establishment is an infrequent occurrence (Bowman and Panton 1993*c*). These facts support Herbert's (1929) suggestion that *E. tetrodonta*'s distribution is relictual. A number of other north Australian eucalypt species also have small populations separated by distances greater than 100 km, and these too may be relictual distributions (Chippendale and Wolf 1981; Bowman *et al.* 1993).

The faunal relationships of *Eucalyptus* also reflect a long evolutionary history. The northern Australia *Eucalyptus* savannas have an extraordinarily rich ant fauna that is different from the relatively impoverished ant assemblages specialised to monsoon rainforest habitats (Andersen 1991; Reichel and Andersen 1996). There is evidence of co-evolution of both vertebrate and invertebrate herbivores with *Eucalyptus* species (Freeland and Winter 1975; Smith and Ganzhorn 1996; Landsberg and Cork 1997). When *Eucalyptus* are grown as exotics free from their co-evolved insect herbivores, they show a marked growth response and often develop denser canopies (Pryor 1976). The dietary specialisation of rat kangaroos to hypogenous fungi that are abundant in *Eucalyptus* forest, particularly on dry and infertile sites, is also consistent with the view of long-term evolutionary relationships between vertebrates and *Eucalyptus*-dominated ecosystems (Johnson 1996). Johnson notes that rat kangaroos are 'exceptional among small mammals in having an enlarged forestomach region in which microbial fermentation of poorly digestible foods can take place, and this digestive strategy apparently allows them to specialise on hypogenous fungi'. The fossil record from Riversleigh south of the Gulf of Carpentaria suggests that a complex assemblage of rainforest mammals gave way to savanna-adapted species (such as grass eating kangaroos) in the late Tertiary (Archer *et al.* 1989). The adaptation of rainforest-dwelling faunal taxa to *Eucalyptus* communities is well exemplified by the radiation of the honeyeaters, largely nectarivorous birds in the family Meliphagidae with many species endemic to *Eucalyptus* forests and woodlands (Figure 11.16). Cladistic analyses of allozyme data show that the majority of species that now occur in *Eucalyptus* habitats are derived from rainforest-adapted taxa (Christidis and Schodde 1993). They speculate that the diversification of honeyeaters is a response to broad-scale environmental change in the late Tertiary.

Bond and van Wilgen (1996) argued that if savannas were of recent origin, woody taxa should be shared between forests and savannas. If savannas are of great antiquity, separate floras should have evolved to produce distinct savanna and rainforest elements including endemic savanna genera. Applying this logic, it is clear that the enormous diversity of *Eucalyptus* vegetation and the profound differences between *Eucalyptus* and rainforest vegetation reflects an evolutionary divergence that occurred over a period in the order of 10^6 or 10^7 years. Indeed, Hopper (1979) concluded that the great diversification and endemicity of the southwest Western Australia flora, including 163 *Eucalyptus* species, arose since the late Tertiary in response to long-term environmental changes. He suggested that these included: (i) late Cainozoic aridification that resulted in the region's

Figure 11.16
The Yellow-plumed
Honeyeater
*(*Lichenostomus ornatus*)*
inhabits low Eucalyptus
woodlands (mallee) in
semi-arid southern
Australia. (Photograph:
Don Franklin.)

biogeographic isolation from eastern Australia; (ii) tectonic stability and associated deep weathering leading to soil infertility; and (iii) a mosaic of edaphic
conditions related to the interplay of climatic fluctuations and geomorphologic
evolution during the late Cainozoic.

The above evidence clearly makes the proposition that Aboriginal burning
caused the continent-wide speciation of eucalypts and associated plants and
animals untenable. It is more likely that the diversity of the non-rainforest
vegetation reflects a long evolutionary history. This point was ably put by Nix
(1982), who wrote that 'the major radiation and differentiation of the Australian
biota took place long before the Quaternary and indeed, for many groups, far
back in the Tertiary. Even at the level of speciation, the Quaternary climatic
fluctuations may ultimately prove to have had no more than a 'cosmetic' effect
upon already existing taxa and their major influence may have been one of
differential extinction'. It is interesting to note that during this century there has

been a conceptual shift in thinking about rainforest and *Eucalyptus*. Early in this century *Eucalyptus* was considered an ancient group (Herbert 1929; Wood 1959) and rainforests as recent colonisers (Brough *et al.* 1924). More recently the view that rainforests were autochthonous and subsequently gave rise to non-rainforest species such as eucalypts has gained ascendancy (Webb *et al.* 1986). I suggest that, in the battle to make this important conceptual shift, there has been an unintended belittling of the great antiquity of non-rainforest species like *Eucalyptus*. Certainly, the bias against xeric vegetation in the fossil record has contributed to the view that eucalypts are not ancient.

Conclusion

There is no doubt that fire was an indispensable tool in Aboriginal economies. Equally, there is no doubt that landscape burning by Aborigines extended the range of fire-adapted species such as eucalypts in environments that would otherwise support luxuriant rainforest. However, Aboriginal burning at the beginning of the Holocene was unable to arrest or reverse the climatically driven expansion of rainforest. Thus, Aboriginal fire usage has had a substantial, but not absolute imprint on Australian biogeography (Jackson 1968; Smith and Guyer 1983). It appears however, that the only significant evolutionary impact of Aboriginal landscape burning may have been the extinction of some fire-sensitive species of plants and animals dependent on large tracts of infrequently-burnt habitat, and that these extinctions probably occurred during periods of climatic stress (Nix 1982; Kershaw 1984; Hill and Read 1987; Macphail *et al.* 1993). It is improbable that Aboriginal use of fire triggered the major diversification of the non-rainforest biota because their period of residence in Australia has not been sufficiently long to have caused such a major evolutionary event. This conclusion begs the following: what caused evolutionary divergence of rainforest and non-rainforest vegetation? This is the subject of the following chapter.

12

The fire theory V. Aridity and the evolution of flammable forests

In this book, I have sought to establish that the boundaries between rainforest and non-rainforest vegetation are determined by the frequency of wildfire. This generalisation holds across a latitudinal gradient from 12° S in the monsoon tropics to 44° S in Tasmania and includes all rainforest environments on the east coast of Australia. Indeed, in arid Australia fire also delimits the boundaries of small patches of woody vegetation that have taxonomic links to 'rainforest', such as *Callitris*-dominated communities (Bowman 1992*b*; Bowman and Latz 1993). The importance of fire in dichotomising the ecology of Australian forests was clearly enunciated by Webb (1968). He wrote that it is possible to 'regard the Australian forest flora as composed of species ranging from extremely fire-sensitive to fire tolerant, which have been sifted by fire as an evolutionary factor to produce two major classes of vegetation: one whose existence depends on protection from fire, and the other which is able to survive or regenerate in different degrees after burning'. The purpose of this chapter is to explore the historical reasons for this ecological divergence. First, I will briefly describe the environmental history of Australian vegetation by reference to the abiotic and biotic evidence. I will then critically consider evolutionary models that account for the diaspora of the habitat islands of fire-sensitive rainforest in vast tracts of flammable forests.

Environmental change in Cainozoic Australia

It is widely accepted that during the last 60 million years, Australia has been transformed progressively from a predominantly humid to a predominantly arid continent (Figure 12.1). Kemp (1978) suggested that in the early Tertiary, Australia had a humid climate because rain-bearing westerly winds were able to

Figure 12.1
Arid landscapes such as the Breaden Hills in the Gibson Desert dominate much of the Australian land mass. (Photograph: Don Franklin.)

penetrate the interior of the continent. At that time, the continent was some 20 degrees further south than now. Since the early Tertiary, the continent has drifted into the 'horse-latitudes' where the drier subtropical high pressure systems predominate. The changed geography and oceanography of the southern hemisphere associated with continental drift is also thought to have been a major contributor to global climatic change. The opening of the Drake Passage between Antarctic and South America in the mid Tertiary, and the consequent development of the circumpolar current, may have limited heat exchange between equatorial and polar ocean bodies (Flohn 1973). It has also been suggested that global cooling in the Miocene was caused by widespread tectonic activity and associated increased chemical weathering that reduced the concentration of atmospheric CO_2 (Raymo and Ruddiman 1992). Planetary cooling and the concurrent uplift of the Transantarctic Mountains triggered the formation of the Antarctic ice cap (Singh 1988). The formation of the Antarctic ice cap and changes in the geographic position of the continents caused a steepening of the thermal gradient between the South Pole and the equator. This in turn resulted in the intensification and northward displacement of both the westerly wind systems and the subtropical high pressure systems (Flohn 1973; Bowler 1982). Bowler suggested that Miocene aridification of Australia occurred progressively

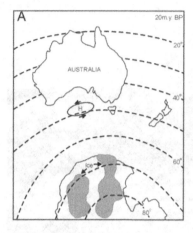

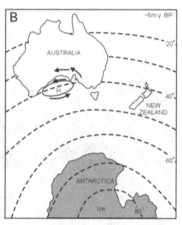

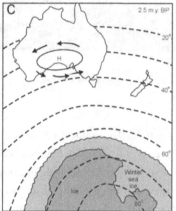

Figure 12.2

Schematic model of the interaction between continental drift, the formation of the Antarctic icesheet and the northward migration and intensification of the subtropical high pressure systems. (A) At around 20 million years ago, mountain glaciers formed in Antarctica but seas were warm to about latitude 50° S. During winter, the high pressure cells moved to the south of Australia. (B) By around 6 million years ago, an icesheet had developed on the Antarctic continent, with associated cooling of sea surface temperatures. The northward movement of intensified high-pressure cells brought aridity to the central coast of southern Australia. (C) At around 2.5 million years ago, winter sea ice formed around the Antarctic continent and the intensified high pressure systems brought aridity to central Australia. Adapted from Bowler (1982).

from the south to the north of the continent because the subtropical high pressure system overtook the slower drift northwards of the Australian continent (Figure 12.2).

Quaternary glaciers and associated periglacial conditions were confined to the highlands of southeastern Australia and Tasmania. Estimated average air temperatures in Central Australia were around 9 °C lower than present at the height of the last glaciation (Miller *et al.* 1997). This cold climate might explain the current dominance of *Acacia* over *Eucalyptus* in central Australia (Bowman and Connors 1996). Throughout eastern Australia, low temperatures during the height of the last glacial depressed the tree line in mountainous areas in eastern Australia, and there was a corresponding expansion of Australia's restricted

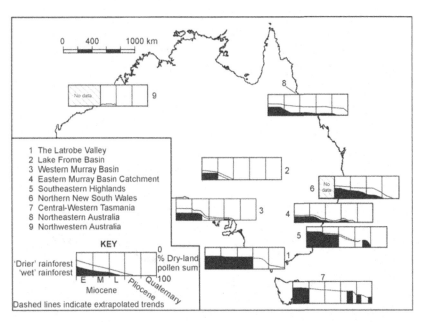

Figure 12.3
Pictorial summary of the pollen evidence for the continent-wide decline of rainforest during the Cainozoic. Adapted from Kershaw et al. (1994) and reprinted with the permission of Cambridge University Press.

alpine vegetation (Hope 1994b). Dune fields and lake sediments across the continent attest to the recurrent fluctuation of arid and humid climates driven by the Quaternary glacial cycles (Bowler 1982; Wasson 1986; Chappell 1991). For example, between 30 000 to 50 000 years ago, Lake Woods (latitude 17° 45' S) in the Northern Territory was ten times its current size (5476 km² versus 423 km²) (Bowler 1981). At the height of the last glacial about 20 000 years ago, extensive dune fields on its shore attest to conditions more arid than present.

Pollen evidence suggests that during the early Tertiary the Australian continent was covered in humid rainforest often dominated by conifers and *Nothofagus* taxa that are now restricted to the extra-Australian tropics. In the mid Tertiary, the mesic rainforests were replaced by drought-adapted rainforest typically dominated by Araucariaceae, and non-rainforest vegetation (Martin 1994). This transition apparently began in central Australia and spread progressively to the coastal margins of the continent (Truswell and Harris 1982). By the late Quaternary, non-rainforest vegetation had almost completely replaced the rainforests. Kershaw *et al.* (1994) provided a pictorial summary of the transition from rainforest to non-rainforest vegetation during the Cainozoic (Figures 12.3 and 12.4). It must be conceded that these patterns are based on a limited

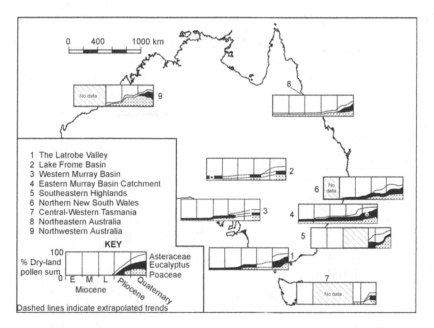

Figure 12.4
Pictorial summary of the pollen evidence for the continent-wide increase of selected non-rainforest taxa during the Cainozoic. Adapted from Kershaw et al. (1994) and reprinted with the permission of Cambridge University Press.

geographic coverage of pollen sites, with especially poor coverage for northern Australia (Harms *et al.* 1980). Despite the poor fossil record for most of Australia, the distribution of numerous animal and plant species is consistent with the concept of previous humid climates. The biogeography of the tree genus *Ilex* and the palm genus *Livistona* illustrates this point.

There is only one *Ilex* species extant in Australia, *Ilex arnhemensis*, and it is restricted to localised moist sites within northern Australia (Figure 12.5). The hypothesis that this species is a Gondwanic relict from humid early-Tertiary climates is supported by pollen evidence, which shows that *Ilex* had a continent-wide distribution in the early Tertiary (Martin 1977), and by the occurrence of *Ilex* species on the islands of New Caledonia and New Guinea. However, birds disperse *I. arnhemensis* so it is possible, though unlikely, that *Ilex* invaded (or re-invaded) Australia from a source area such as New Guinea (Martin 1977).

Livistona is a diverse genus of fan palms found in Australia and southeast Asia (Figure 12.6). The centre of diversity of the genus is northern Australia, where most taxa are regional endemics with disjunct populations. Some species are endemic to rainforest and others endemic to non-rainforest vegetation, and a

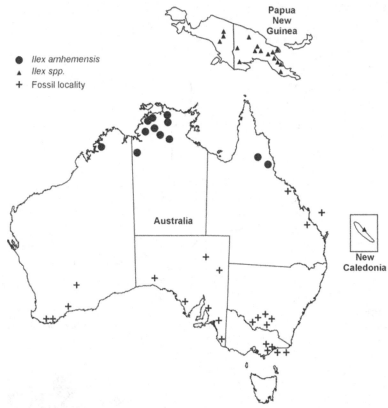

Figure 12.5

Change in the distribution of Ilex *in Australia based on herbarium and fossil pollen records. The current distribution of* Ilex *in New Guinea and New Caledonia is also indicated. Adapted from Martin (1977).*

few occur in localised environments such as springs, soaks or sheltered canyons in arid Australia (Latz 1975; Hnatiuk 1977; Humphreys *et al.* 1990; Russell-Smith 1991) (Figure 12.7). There are no known fossils of *Livistona* in Australia. Dowe (1995) suggested that the current distribution patterns and taxonomic variability of Australian *Livistona* species are the product of vicariant speciation in habitats fragmented by Tertiary aridity. In contrast, Dowe also suggested that the less taxonomically variable extra-Australian species may have evolved in environments less disturbed by climatic change.

Localised populations of *Livistona* may have arisen following long-distance dispersal events rather than habitat fragmentation (Beadle 1981; Dransfield 1981; Martin 1982), and in any particular case it is difficult to partition out these effects (Barlow and Hyland 1988). However, the large number of plant and animal

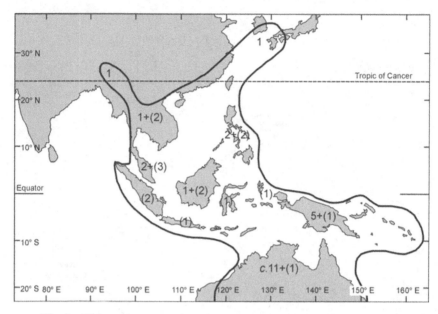

Figure 12.6
The distribution and number of endemic and non-endemic species (in parentheses) of Livistona *species. Adapted from Dransfield (1981), by permission of Oxford University Press.*

Figure 12.7
An oasis in central Australia, dominated by the palm Livistona mariae. *(© Murray Fagg, Australian National Botanical Gardens.)*

species that have disjunct populations in arid Australia makes the hypothesis of long-distance dispersal difficult to sustain as a generality. Latz (1975) noted that, of the 333 vascular plant species that occur in the Palm Valley region of central Australia where *Livistona mariae* occurs, about 10% are rare or restricted in central Australia. The restriction of worm species to a stand of *Livistona mariae* at Palm Valley reinforces the relictual distribution hypothesis (Dyne 1991). Indeed, many canyons in the central Australian mountains support highly disjunct populations of plant and animal species otherwise known from the north, west and east coasts. These species include the shrub *Trema aspera* and the cycad *Macrozamia macdonnelli* (Chippendale 1963), the tree frog *Hyla rubella* and the non-aestivating fish *Fluvialosa richardsoni* (Keast 1959).

Webb *et al.* (1986) rejected the idea that long-distance dispersal could explain the numerous disjunct populations of many Australian rainforest plants. They believed that the predictable spatial repetition of groups of plants, including numerous endemic species, could only be explained by habitat fragmentation. This was especially so because many species have very limited capacity to disperse their propagules. They argued that for these patterns to arise by long-distance dispersal demanded the improbability of 'synchronisation of dispersal in space and time'. Conversely, a number of authors have noted the low rates of plant endemicity in many rainforest patches in northern Australia, and the corresponding high vagility of the monsoon rainforest flora (Barlow and Hyland 1988; Russell-Smith and Lee 1992; Fensham 1995). Russell-Smith and Lee concluded that most monsoon rainforest plants species in the Northern Territory were highly vagile given: (a) their wide geographic range throughout Australia and southeast Asia including geologically recent islands; (b) their frequent occurrence on Holocene landforms; and (c) diaspore characteristics that favoured dispersal by winged vertebrates or wind. The vagility of many plant species' propagules would render the highly restricted current distribution of monsoon forests puzzling but for the overwhelming influence of fire in limiting the distribution of this type of rainforest (Fensham 1995).

Webb and Tracey (1981) listed a number of rainforest genera, including *Livistona*, that have congeners in non-rainforest vegetation, and called the latter 'interspersive elements' (Figure 12.8 and Table 12.1). They assumed that the rainforest taxa were ancestral and that aridification had triggered the subsequent evolution of non-rainforest species. The limited number of phylogenetic studies of both animal and plant groups have not supported Webb and Tracey's conjecture. For example, cladistic analysis of the morphology of the plant genus *Lomatia* in the family Proteaceae did not support the idea that species that occur

Figure 12.8
The palm Livistona
inermis *growing on an
exposed sandstone outcrop
in Arnhemland, Northern
Territory. The species
typically occurs in open and
often rocky situations on
skeletal soils. (Photograph:
Don Franklin.)*

in *Eucalyptus* forests are necessarily derived from rainforest species (Weston and
Crisp 1994). Similarly, although tree kangaroos (*Dendrolagus* spp.) are rainforest
specialists, cladistic analysis of albumin demonstrated that their nearest relatives
are rock wallabies (*Petrogale* spp.) that have adapted to harsh, rocky habitats
throughout Australia (Baverstock *et al.* 1989). There is clear evidence that tree
kangaroos arose from kangaroo species adapted to a terrestrial lifestyle (Flannery
et al. 1995). It is conceivable that these ancestors occurred in non-rainforest
habitats, given that the Miocene diversification of the macropods (Flannery
1989) coincided with the breakdown of Australian rainforest. Thus, the relationship between rainforest and non-rainforest congeners is not necessarily unidirec-
tional. Rather, there is a real possibility that ancestral species have moved
between non-rainforests and rainforest environments in the course of evolutionary time.

   In addition to aridification, global cooling and the seasonality of climate have

TABLE 12.1. Examples of rainforest tree genera with species interspersed in *Eucalyptus* savannas in northern Australia

Family	Rainforest genus	Species in open woodland
Anarcardiaceae	*Buchanania*	*B. obovata*
Apocynaceae	*Wrightia*	*W. saligna*
Boraginaceae	*Ehretia*	*E. saligna*
Burseraceae	*Canarium*	*C. australianum*
Capparidaceae	*Capparis*	*C. lasiantha*
Celastraceae	*Siphonodon*	*S. pendulum*
Combretaceae	*Terminalia*	*T. canescens*
Euphorbiaceae	*Excoecaria*	*E. parvifolia*
Lecythidaceae	*Barringtonia*	*B. gracilis*
Meliaceae	*Owenia*	*O. acidula*
Mimosaceae	*Albizia*	*A. basaltica*
Moraceae	*Ficus*	*F. opposita*
Myrtaceae	*Syzygium*	*S. eucalpytoides*
Oleaceae	*Notelaea*	*N. microcarpa*
Palmae	*Livistona*	*L. muelleri*
Pittosporaceae	*Pittosporum*	*P. phylliraeoides*
Proteaceae	*Grevillea*	*G. parallela*
Rhamnaceae	*Alphitonia*	*A. excelsa*
Rubiaceae	*Gardenia*	*G. vilhelmii*
Sapindaceae	*Atalaya*	*A. hemiglauca*
Sapotaceae	*Planchonella*	*P. pohlmaniana* var. *vestita*
Solanaceae	*Duboisia*	*D. hopwoodii*
Sterculiaceae	*Brachychiton*	*B. diversifolium*
Verbenaceae	*Clerodendrum*	*C. cunninghamii*

Adapted from Webb and Tracey (1981).

also influenced the biogeography of Australian rainforest. There is evidence of an interaction between continental drift and global cooling in palaeo-temperature reconstructions based on oxygen-isotope records in deep-sea cores (Bowler 1982; Truswell 1993). Deep-sea cores from areas of the Southern Ocean near Antarctica that have experienced little northward drift provide evidence that sea surface temperatures decreased progressively during the Cainozoic. In contrast,

TABLE 12.2. Distribution of leaf size classes of trees and vines on an elevation gradient in the Barrington Tops area, New South Wales

Mean altitude above sea level (m)	Mesophylls (%)	Notophylls (%)	Microphylls (%)
999	0	22	78
948	0	17	83
920	0	31	69
891	2	29	69
836	5	36	59

Adapted from Turner (1976).

deep-sea cores from northeastern Australia show that northward continental drift offset the effect of global cooling on sea surface temperatures since the mid Tertiary. Drift of the Australian continent may have affected tree architecture and thus vegetation structure because of changes in day length and sun angle (Kuuluvainen 1992). Continental drift may have also reduced the competitive advantage of species adapted to survive dark winters (Read and Francis 1992), a possible example of which is *Nothofagus gunnii*, a species which is deciduous in winter and now restricted to the mountains of Tasmania (Hill 1994*b*). Microphylly is a distinguishing feature of rainforest trees at both high latitudes and high altitudes where winters are cold and summers cool (Webb 1959, 1968; Webb *et al.* 1986; Nix 1991; Turner 1976; Hill and Read 1987) (Table 12.2). Small leaves are thought to have evolved in response to the global cooling of the late Tertiary, and may have contributed to the extinction of plants with 'warm temperate affinities' in Tasmania (Macphail *et al.* 1993). Macrofossils demonstrate that *Nothofagus cunninghamii* is derived from an ancestor with larger leaves (Figure 12.9) (Hill 1983; Hill and Read 1987). Some rainforest trees with small leaves are more frost tolerant and have higher photosynthetic rates at low temperatures than larger-leaved congeners (Hill *et al.* 1988; Read and Hill 1989; Read and Busby 1990). However, reduction in leaf size may not be solely a response to low temperatures. *N. cunninghamii* is relatively drought-tolerant (Howard 1973*b*). Further, it is curious that the leaf sizes of *Eucalyptus* species do not show a similar relationship between latitude and altitude. More comparative ecophysiological studies, particular of the water relations of small- and large-leaved trees are required to advance these questions.

Nix (1982) has suggested that 'increasing seasonality of both thermal and water regimes throughout the Tertiary have been the primary factor in the

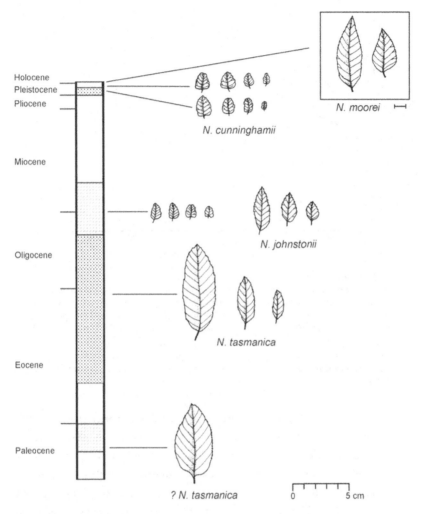

Figure 12.9
Change in leaf size and shape of Nothofagus *taxa in the subgenus* Lophozonia *over time in southeastern Australia. The shaded section on the stratigraphic column shows the approximate temporal range of each species. Adapted from Hill (1994c) and reprinted with the permission of Cambridge University Press.*

evolution of the Australian biota'. Pronounced rainfall seasonality in the late Tertiary may have resulted in the restriction of *Nothofagus moorei* to the subtropics and caused the Australian extinction of extant New Guinean *Nothofagus* taxa (Kershaw 1984; Read and Hill 1989; Read and Hope 1989; Read 1990). Currently, most Australian rainforests occur in climates that have seasonal rainfall and, with the exception of *Nothofagus*-dominated microphyll/nanophyll

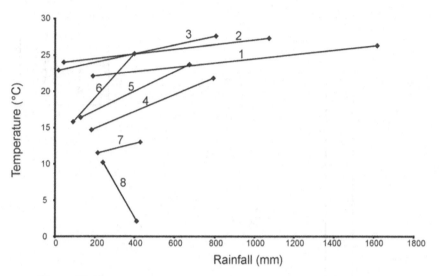

Figure 12.10

Estimated range of mean temperatures and mean precipitation for the wettest and driest quarters for eight groups of Webb's (1968) physiognomic Australian rainforest. Group 1: evergreen mesophyll vine forest (CMVF, MVF); 2: semi-deciduous notophyll/mesophyll vine forest (SDMVF, SDNVF); 3: deciduous microphyll vine thicket (DVT); 4: evergreen notophyll vine forest (CNVF, NVF); 5: evergreen notophyll/microphyll vine forest (ANVF, AMVF); 6: semi-evergreen microphyll vine thicket (SEVT); 7: evergreen microphyll fern forest (MFF); 8: evergreen microphyll/nanophyll moss forest/thicket (MMF, NMT). Data derived from Nix (1991).

forests, the wettest quarter is warmer than the driest quarter (Figure 12.10). Webb (1968) noted that deciduousness is a common trait in most Australian rainforests, suggesting that this phenological pattern is an adaptation to the seasonality of rainfall. These hypotheses are difficult to evaluate because the details of the climates of the Cainozoic are poorly understood, in large part because of the difficulties in developing accurate chronologies of sediments that are usually barren of fossils. Nonetheless, geomorphological evidence such as deeply weathered sediments led Bowler (1982) to conclude that Pliocene climates in southern inland Australia were characterised by summer rain and winter aridity somewhat analogous to the prevailing climate in northern Australia (Figure 12.11).

How long northwestern Australia has had a seasonally arid climate is unknown. The only known plant macrofossil deposit in northern Australia has been interpreted as indicating that seasonal aridity has been a feature of the climate since the early Tertiary (Pole and Bowman 1996; Pole 1998). However, a vertebrate fossil assemblage from the late Oligocene to mid Miocene, preserved

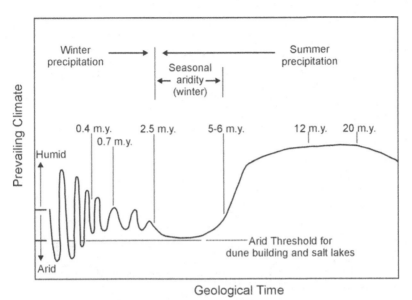

Figure 12.11

Pictorial description of the possible rate and nature of the transition from humid to arid climates in southern inland Australia since the Miocene. Adapted from Bowler (1982).

in limestone at Riversleigh near the Gulf of Carpentaria, has been interpreted as indicating the occurrence of humid tropical rainforest (Archer *et al.* 1989, 1994). The fossils included a 'high diversity of sympatric arboreal folivorous marsupials' such as ringtail possums, and other vertebrate groups that are currently restricted to humid rainforests. Archer *et al.* suggested that mid-Tertiary aridification caused the replacement of rainforest mammals by grazing mammals at Riversleigh. Vertebrate fossil assemblages of mid Tertiary age from central Australia have also been interpreted as being from a seasonally arid environment (Murray and Megirian 1992). Global cooling and drying in the mid Miocene is thought to have stimulated the co-evolution of African grasslands and large mammalian herbivores (Retallack 1992). The biogeography of the Orange Horse-shoe Bat *Rhinonycteris aurantius* is consistent with the hypothesis that seasonally arid climates developed in the mid Tertiary in northern Australia. This bat has a fragmentary distribution throughout northern Australia (Figure 12.12). Fossils of the genus are known from Oligocene–Miocene deposits at Riversleigh (Archer *et al.* 1989), a region where the Orange Horse-shoe Bat is not currently known. Because this species is intolerant of the 'low' air temperatures characteristic of the tropical dry season nights, colonies are restricted to cave environments that are hot (28–32°C) and humid (85–100% relative humidity)

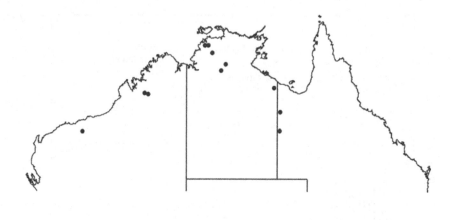

• Colonies

Figure 12.12
Distribution of colonies of the Australian endemic Orange Horse-shoe Bat Rhinonycteris
aurantius. *Adapted from Churchill (1991).*

(Churchill 1991). This species is unable to survive current climatic conditions
and persists only in thermal refugia.

Another interpretation of the Riversleigh deposit is that the rainforest was
restricted to a watercourse, a habitat also known as 'gallery forest', and that the
regional climate was 'relatively dry, perhaps semi-arid', supporting large ex-
panses of non-rainforest vegetation (Megirian 1992). Bowman and Woinarski
(1994) noted that in South American savanna zones, gallery rainforests support
the majority of the mammalian fauna and the adjacent savannas few species,
while the reverse is true in Australian savannas, possibly because of the dominance of *Melaleuca* on most river lines. The diversification of *Melaleuca* parallel-
ed the radiation of *Eucalyptus*. They speculated that *Melaleuca* may have replaced
floristically diverse gallery rainforests. If this was the case, it might explain the
mid-Tertiary extinction of many vertebrates at Riversleigh.

Sometime in the late Pleistocene, a range of marsupial megafauna that were
adapted to open conditions became extinct, and dwarfing of surviving lineages
occurred (Horton 1984; Flannery 1990*a*; Archer *et al.* 1994). It is hotly debated as
to whether the Pleistocene megafauna became extinct because of climatic change,
Aboriginal over-hunting or a combination of both factors (Horton 1984; Flannery 1990*b*). Available evidence has not permitted resolution of the debate,
although recent mathematical modelling casts doubt on the idea that megafaunal
extinction was solely caused by Aboriginal over-hunting (Choquenot and Bowman 1998).

In summary, climate change and particularly aridification through the Cainozoic contributed to the local and global extinction of many plant and animal species and led to the fragmented and otherwise puzzling distributions of rainforest patches and numerous individual plant and animal species in Australia. However, the demonstrable tolerance of many Australian rainforest tree species to a wide range of climates (Chapters 6 and 7) suggests that climate change alone cannot explain the transition from rainforest to non-rainforest.

The fire–aridity interaction

The drought tolerance of many Australian rainforest tree species (Chapter 6) could have insulated them from the consequences of Cainozoic aridification. Kershaw (1981) noted the paradoxical nature of the hypothesis that dry Quaternary climates caused the contraction of drought-tolerant rainforest taxa. He wrote:

there are some vegetation changes which would not be expected under change in climate alone, for example the replacement of Araucaria forests by open forests or woodland between 38 000 and 26 000 ago on the Atherton Tableland. Although there is evidence of a reduction in precipitation at this time, drier conditions alone would have tended to cause a change to a less complex closed forest community such as Deciduous Vine thicket which exists in isolated patches, particularly on basaltic soils down to about the 500 mm isohyet – rather than eucalypt woodland.

Similarly, it is improbable that the heavily fragmented distribution of *Callitris* in Australia could be solely the work of aridification. The genus *Callitris* is composed of xerophytic species that were abundant during the arid phase of the last ice age (Attiwill and Clayton-Greene 1984; Singh and Luly 1991; Bowman and Harris 1995). Some authors have argued that resolution of this paradox lies with the effects of Aboriginal landscape burning in the late Pleistocene (Kershaw 1981, 1986; Singh and Geissler 1985; Kershaw *et al.* 1993). They suggest that Aborigines caused a shift from vegetation that was fire-sensitive to fire-adapted vegetation. In the previous chapter, I rejected this hypothesis, firstly because there is no clear evidence of a massive impact of Aboriginal landscape burning in the late Pleistocene and secondly because there is insufficient time for the massive diversification of the non-rainforest biota to have occurred since Aboriginal colonisation.

Martin (1994) has advanced the alternative hypothesis that 'the decline in rainforest was caused by climate change', not by Aborigines, because 'fire had become an integral part of the environment in the late Miocene'. Martin's

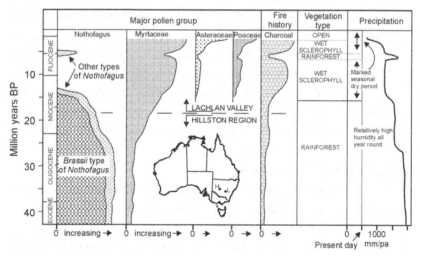

Figure 12.13

Diagrammatic summary of changes in the abundance of the major pollen taxa and microscopic charcoal particles from a composite of pollen cores in the Lachlan valley and Hillston regions of central New South Wales. Also indicated are vegetation types and prevailing climates inferred from the fossil pollen assemblages. Adapted from Martin (1987).

hypothesis is based on palynological interpretation of a composite pollen diagram based on several well-separated cores in the Lachlan River Valley in central New South Wales (Figure 12.13). Martin (1987) concluded that, before the Miocene, *Nothofagus* dominated diverse rainforests for which there is no extant analogue. The dominant *Nothofagus* pollen was of the *brassii* type produced by species that are now restricted to the extra-Australian tropics. Pollen of the *N. fusca* type (now represented in Australia by *N. gunnii*) and *N. menziesii* type (now represented in Australia by *N. cunninghamii* and N. *moorei*) was also present in the cores. Throughout the first half of the Miocene, *brassii* pollen steadily declined until it became locally extinct. The *Nothofagus* pollen assemblage was replaced by Myrtaceae pollen associated with high concentrations of microscopic charcoal particles. Martin argued that this assemblage is ecologically analogous to a wet *Eucalyptus* sclerophyll forest, and that fires caused the replacement of the *Nothofagus* rainforest. In the early Pliocene, *Nothofagus fusca* and *menziesii* pollen types appeared in conjunction with gymnosperm pollen and a low concentration of charcoal particles. By the mid–late Pliocene, Myrtaceae pollen was again dominant, with a corresponding increase in charcoal particles. The Pleistocene pollen assemblage is rich in Asteraceae and Poaceae pollen and is thought to reflect the expansion of grasslands. Martin suggested that the changes in pollen assemblages were primarily driven by global climate change and particularly the aridity that followed the formation of the Antarctic ice cap. It was her view that fire should be seen as a direct consequence of a drier climate. Her interpretation is consistent with charcoal particles and pollen evidence for fires in Miocene sediments in the Latrobe Valley region of eastern Victoria (Kershaw *et al.* 1994).

The above description of the shift from rainforest to *Eucalyptus* forests is broad-brush and indicative of a trend. A deficiency of this description is that it provides no explanation for why fire should have become an important factor in the Tertiary environment. Some hybrid hypotheses have been proposed. One suggestion is that the transition from rainforest to non-rainforest vegetation was caused by late-Tertiary aridity and that this trend was accelerated in the late Quaternary by the increased frequency of fires associated with human colonisation (Kemp 1981; Truswell 1993). Smith and Guyer (1983) hypothesised that *Eucalyptus* originally evolved as a 'pioneer species' or 'biological nomad' in rainforest environments by exploiting natural disturbances such as land slips and tree fall gaps. They believed that Aboriginal burning favoured their dominance but was not the cause of their dependence on disturbance. The conservative interpretation of these hybrid hypotheses is that Aboriginal landscape burning rearranged existing species and vegetation types in the landscape. A more radical interpretation is the idea that Aboriginal landscape burning triggered a major diversification of non-rainforest species. I argued in the last chapter that this latter view is not supported by a great mass of diverse biogeographic facts. However, rejection of the hypothesis that Aboriginal landscape burning forced the evolution of the Australian biota begs the following questions. What caused the evolution of flammable vegetation? Did mid-Tertiary aridity simply increase the frequency of fire and cause the evolution of fire-adapted vegetation? Alternatively, did the evolution of fire-adapted vegetation increase the frequency of fire?

Evolution of flammable forests

The refined interdependence of fire and non-rainforest vegetation in Australia has been widely regarded as the work of evolution (Mount 1964; Jackson 1968). Such interdependence is often illustrated by the fire ecology of *Eucalyptus* species such as *E. regnans* and *E. grandis*, which occur in rainforest environments. These eucalypts retain seeds in woody capsules and the seeds are released *en masse* following fire (Cremer 1965; Cremer and Mount 1965; Ashton 1981*a*). Establishment of seedlings of these species on unburnt sites is extremely rare (Fraser and Vickery 1939; Gilbert 1959; Mount 1979; Ellis *et al.* 1980). Furthermore, seedling growth is suppressed on unburnt soil but is rapid on sterilised soil as typically occurs following fire (Pryor 1956; Cremer 1962; Florence and Crocker 1962; Loneragan and Loneragan 1964; Renbuss *et al.* 1972; Ashton 1976*b*; Ashton and Willis 1982; Ellis and Pennington 1992; Bowman and Fensham 1995). All these features are inconsistent with Smith and Guyer's (1983) suggestion that such eucalypts are of recent derivation from 'secondary' rainforest trees that opportunistically colonise rainforest gaps. Indeed, *Eucalyptus* species occupying rainforest environments have been described as 'transient fire weeds' (Cremer 1960).

Despite the vast body of field evidence, evolutionary theories that account for the interdependence of fire and some tree species remain 'embryonic' (Bond and van Wilgen 1996). Research has focused on two aspects of this problem: (a) how plants evolve traits that enable them to recover from fire; and (b) how plants evolve traits that increase their flammability. These topics are discussed below with reference to *Eucalyptus* and rainforest.

Mechanisms to recover from fire

Many authors have stressed that *Eucalyptus* and other non-rainforest species can recover from fire because they have evolved specialised morphological and anatomical features including lignotubers, thick fibrous bark, epicormic buds and hard woody capsules (Jackson 1968; Ashton 1981*a*) (Figure 12.14). However, there is no evidence that these features developed *de novo* in order to tolerate recurrent fires. This point was appreciated by Jarrett and Petrie (1929), who wrote that 'every character by which the Australian vegetation recovers itself after the onslaught of fire seems unquestionably to have been developed for other purposes'. Second, many rainforest trees also possess these alleged 'fire-tolerating' adaptations. For example, Johnston and Lacey (1983) and Lacey and Jahnke (1984) have shown that lignotubers are common in a diverse range of rainforest tree lineages including families such as Myrtaceae, Casuarinaceae, Proteaceae, Tremandraceae, Leguminosae, Sterculiaceae and Dilleniaceae. They speculated that the lignotuber has an ancient origin and is probably 'not (or only slightly) related to fire or drought'. Of course, the fact that rainforest species have characteristics such as lignotubers does not preclude the possibility that natural selection has enhanced and perfected such features in order to allow genotypes to survive in environments subjected to frequent fires. An example of a rainforest tree species that could potentially give rise to a fire-adapted species is *Allosyncarpia ternata*. *A. ternata* is thought to be a close relative of *Eucalyptus*, and is sometimes used as an out-group in cladistic analyses of that genus (Udovicic *et al.* 1995). *A. ternata* forests are being transformed into *Eucalyptus* savanna

Figure 12.14
Epicormic buds on the trunk of a Eucalyptus *tree 11 months after being burnt by a severe wildfire, southeastern Australia. (© Murray Fagg, Australian National Botanical Gardens.)*

because it is less tolerant of the current regime of frequent high-intensity fires than trees such as *E. tetrodonta* (Bowman 1991*a*, 1994). Nonetheless, *A. ternata* possesses a range of characters that confer some tolerance to fire, including thick fibrous bark and a lignotuber (Fordyce *et al.* 1997*b*). It is possible that limited genetic changes in an ancestral species allied to *A. ternata* may have given rise to more fire-tolerant genotypes which in turn gave rise to *Eucalyptus* species.

Paradoxically, the suppression of vegetative recovery mechanisms may represent an evolutionary strategy to survive recurrent fires (Bond and van Wilgen 1996). A plant that does not develop a lignotuber may be able to sustain higher juvenile growth rates because more carbon can be allocated to the development of foliage and fine roots. *Eucalyptus regnans* is an example of a species employing this strategy (Ashton 1981*a*). Following a fire, *E. regnans* releases seeds stored in woody capsules, a phenological pattern known as 'serotiny'. Ecologically speaking, serotiny is analogous to mast fruiting of trees because both syndromes may have evolved to satiate seed predators. The essential difference between these syndromes is the retention time of the seed and the cue for seed release. There is some evidence that only minor genetic changes are required to cause a tree to change from being non-serotinous to serotinous (Bond and van Wilgen 1996). In this context, it is interesting to note that there is considerable variation in the

Figure 12.15
Damage to robust woody fruits of the monsoon tropical tree Eucalyptus ptychocarpa *by Red-tailed Black Cockatoo* Calyptorhynchus banksii. *It is possible that there is a biological 'arms race' between the fruits of* Eucalyptus *and the beaks of cockatoos. (Photograph: David Bowman.)*

mode of seed release of *Eucalyptus* species. Species such as *E. regnans* that occur in infrequently burnt moist habitats are serotinous. In contrast, most eucalypts in the monsoon tropics, where fire is a frequent occurrence, do not retain seeds in woody fruits once they have matured (Dunlop and Webb 1991; Setterfield and Williams 1996).

Some workers believe that robust woody capsules typical of many *Eucalyptus* species evolved to protect seeds from fire (Mount 1964; Jackson 1968). However, there is little evidence to support this view. Experimental studies with four southern Australian species in the family Myrtaceae led Judd (1994) to conclude that the survival of seeds in woody capsules is a function of flame residence times rather than any special features of woody capsules. An alternative function of woody capsules would appear to be protection from vertebrate seed predators. Vertebrates are a significant cause of seed mortality despite woody capsules (Figure 12.15). For instance, Setterfield and Williams (1996) found that Red-tailed Black Cockatoos (*Calyptorhynchus banksii*), a large bird with an enormously powerful beak, were a significant cause of *E. tetrodonta* and *E. miniata* seed mortality even though these species have heavy woody fruits.

Experiments have shown that under field and nursery conditions, smoke, or solutions containing compounds derived from smoke, stimulate the germination of numerous Australian plant species. These include some species that have low germination rates or remain dormant in the absence of this pre-treatment (Dixon *et al.* 1995; Roche *et al.* 1997*a,b*, 1998). Smoke-induced germination strongly suggests a close evolutionary relationship with fire, but research is also required to determine the effect of smoke on the germination of rainforest species.

In summary, it is clear that non-rainforest species have a range of traits that enable their regeneration following fire. Though the development of evolutionary thinking on the subject is still in its infancy, there are credible explanations as to how these traits have developed. I now consider whether there was a corresponding evolution of characteristics to increase the frequency of fire.

Mechanism to propagate fires

Bond and van Wilgen (1996) noted that 'the evolution of flammability in just a single common species could trigger cascading effects on ecosystem composition by causing an increased occurrence of fires'. Further, they noted that the radiation of species like eucalypts might have been driven by the 'spread of fire-prone vegetation'. Mount (1964) and Jackson (1968) were amongst the first authors to hypothesise that *Eucalyptus* species have evolved characteristics to increase the frequency of fire. Fire-promoting features of *Eucalyptus* are thought to include 'development of fire-brands of bark which encourage spot fires, the flammable oil content of foliage and twigs, heavy litter fall, and open crowns bearing pendulous foliage which would encourage maximum updraught. The fibrous bark of some wet sclerophyll eucalypts will protect the cambium effectively from heat but can carry the fire to the canopy, thereby encouraging development of a crown fire' (Ashton 1981*a*) (Figure 12.16). Mount (1979) believed that the capacity of vegetation to propagate fires meant that an area as large as Tasmania could be burnt in the absence of intervention by humans, even if natural ignitions, such as lightning, were relatively infrequent.

The hypothesis that some vegetation types have evolved characteristics to increase the risk of fire was generalised and popularised by the northern American ecologist Mutch (1970). Mutch argued that 'fire-dependent plant communities burn more readily than non-fire-dependent communities because natural selection has favoured development of characteristics to make them more flammable'. This evolutionary conjecture is flawed when stated in this way for the following four reasons.

Figure 12.16
Highly flammable mallee vegetation, northwestern Victoria. High concentrations of essential oils that have low flash points causes the explosive combustion of the Eucalyptus *crown. (Photograph: Don Franklin.)*

(i)   The argument is inherently circular; flammable vegetation must have characteristics which render it flammable (Snyder 1984; Bowman and Wilson 1988; Troumbis and Trabaud 1989).

(ii)   Traits that cause flammability may arise in response to other selective forces such as herbivory, drought or infertility (Snyder 1984; Troumbis and Trabaud 1989). For instance, some authors believe that the high content of flammable oils in the litter and foliage of many eucalypts is an adaptation that increases the risk of fires (Pompe and Vines 1966; Jackson 1968; Webb 1968; Ashton 1981a). An alternative evolutionary explanation is that the oils are a chemical defence against herbivores (Morrow and Fox 1980, Landsberg and Ohmart 1989).

(iii) Characters that appear to be an adaptation to increase fire risk may have arisen independently of any selective pressure. For example, Ellis (1965) doubted that *Eucalyptus* leaf litter is an evolutionary adaptation to produce fuel for forest fires. Equally, the presence of ribbon bark on Myrtaceous trees in southeast Asian rainforests cannot be interpreted as an adaptation to propagate fires via Mount's (1964) 'spot-fire mechanism' hypothesis, because crown fires do not occur in these forests.

(iv)   The Mutch hypothesis is based on the concept of 'group selection'
because it does not explain why flammability would increase the
reproductive fitness of a highly flammable genotype (Zedler 1995).
Mount (1964) also employed a group selection argument. He wrote 'the
mechanism of spotting caused by these types of bark is of no benefit to
the individual currently alight, but enables the fire to cross major
obstacles such as wet gullies. In Tasmania certain eucalypt species
appear to have been maintained by spot fires on isolated sites
completely surrounded by rainforest'.

Some workers have attempted to refine the Mutch hypothesis in order to
overcome these problems. Buckley (1984), Bond and Midgely (1995) and Possingham *et al.* (1995) argued that fire should be seen as an agent of inter-specific
competition where burning damages competitors but can increase the reproductive success of a flammable species. Bond and Midgely neatly sum this idea up by
commenting wryly that 'flammable plants make dangerous neighbours'. They
posit that the capacity of flammable genotypes to interfere with non-flammable
competitors provides clear evolutionary advantage (Figure 12.17) and used
cellular automata mathematical models to demonstrate that flammable individuals increase their reproductive success by killing less flammable neighbours.
However, this could not occur in plant populations with low densities. Their
modelling also showed that the spread of a trait promoting flammability through
a population could only occur if the trait increased reproductive fitness. They
suggested that this might occur if the flammability trait also conferred other
evolutionary advantages, such as by providing chemical defences against herbivory.

It is impossible to prove conclusively that natural selection has produced fire-
promoting characteristics. There is some correlative evidence consistent with the
view that chemical features of foliage can increase the risk of fire. King and Vines
(1969) compared the flammability of oven-dried leaves from trees from a range
of different Australian rainforest and non-rainforest types. They found that in
general, *Eucalyptus* had more highly combustible foliage than rainforest trees.
They attributed this difference in part to the higher concentration of mineral
elements in rainforest foliage. Dickinson and Kirkpatrick (1985) determined the
energy content and the rate of flame movement in dead and living foliage from a
range of Tasmanian tree species that occur in wet and dry *Eucalyptus* forests.
They showed that species from the frequently burnt dry forests had a greater
range of energy contents (16.0 to 23.5 kJ g^{-1}) than the less frequently burnt mesic

Fire intolerant genotype   Flammable genotype

FIRE

Enhanced flammability increases risk of neighbour mortality

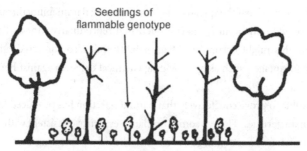

Seedlings of
flammable genotype

Flammable plants exploit gaps created by burning,
especially if they have additional growth advantages

Figure 12.17

Diagrammatic summary of the argument that characters that increase the flammability of a genotype could result in its increased reproductive fitness such that the genes for flammability spread through the ancestral less flammable population. By killing its neighbours, a flammable plant provides increased recruitment opportunities to seeds with the flammable genes. The modelling of Bond and Midgley (1995) suggested that genes that enhance flammability are most likely to dominate a population if they also provide some other additional benefits such as faster growth rates or defences against herbivory. Close packing of individual plants is required to ensure that neighbours can be burnt by a flammable genotype. Adapted from Bond and van Wilgen (1996), with kind permission from Kluwer Academic Publishers.

forests (20.2 to 24.8 kJ g^{-1}). Moisture content strongly influenced the rate of flame spread. Some of the dry forest species had highest rates of flame spread in fresh foliage, while some of the wet forests species had lowest rates of flame spread in fresh foliage. Rates of flame spread in desiccated foliage were comparable for both wet and dry forest species. Consequently, under drought conditions the foliage of wet *Eucalyptus* forests becomes highly flammable.

In contrast to these results, Bowman and Wilson (1988) found that dry monsoon rainforest species had higher leaf energy contents (18.4 to 22.5 kJ g^{-1}) than leaves from adjacent *Eucalyptus* savanna trees (16.2 to 20.0 kJ g^{-1}). In general, rates of flame spread in dead and dry foliage of savanna and monsoon rainforest species were similar. An interesting exception was the dead foliage of the rainforest tree *Drypetes lasiogyna*, which had very high rates of flame spread. Bowman and Wilson concluded that the infrequently burnt monsoon rainforests were better discriminated from the very frequently burnt *Eucalyptus* savanna by grass biomass and fuel density than the flammability of foliage (Table 12.3). Given that grassy fuels are highly flammable, no further adaptations are required to ensure a high frequency of fire in tropical savannas. Bowman and Wilson speculated that 'evolution of fire-promoting foliage occurs where natural ignitions are infrequent'. Natural selection may have changed chemical characteristics in existing foliage to increase flammability. This view is consistent with the observation that *Eucalyptus* species in humid environments have higher concentrations of oils in their foliage than *Eucalyptus* species in the monsoon tropics (Webb 1968).

Synthesis

Rainforests became increasingly susceptible to destruction by fire as the Australian climate dried out during the Tertiary. The severe droughts in southeast Asia in 1982–83 and 1997–98 clearly demonstrate how humid tropical rainforest can be destroyed by fire (Malingreau *et al.* 1985; Brown 1998). Admittedly, the scale of these fires was linked to unsustainable forestry practices, land clearance and human ignitions. However, it is reasonable to assume that over millions of years, arid climates and natural ignitions could cause the destruction of large areas of rainforest. The increasing dominance of the subtropical high pressure system in the Australian climate would have been of particular importance in promoting the spread of fires. Currently, these subtropical highs are responsible for strong, dry winds in the rain-free summer months in southern Australia and the rain-free winter months in northern Australia. The gradual climatic transition would have been an ideal evolutionary theatre for the development of fire-

TABLE 12.3. Density of grass and litter layers in monsoon rainforest and *Eucalyptus* savanna (mean and range). Mann–Whitney U-tests showed that these fuel components were significantly different ($P < 0.01$)

	Community	
Fuel component density	Monsoon rainforest	*Eucalyptus* savanna
Grass density (g m^{-3})	0 (0)	63.8 (37–193)
Litter density (g m^{-3})	35.2 (18–63)	20.4 (11–39)

Adapted from Bowman and Wilson (1988).

adapted vegetation. Once fire-dependent vegetation had developed, the frequency and extent of fires would have been further promoted because of the availability of fuels. The spread of fires would also have been favoured by the subdued topography of most of Australia. Luke and McArthur (1978) estimated that in the 1974–75 fire season, fires caused by lightning burnt almost one-sixth of the Australian continental land mass.

In spite of the evidence to support the view that non-rainforest species developed adaptations to recover from fire and to increase their flammability, it remains unclear how these evolutionary feats were initiated or choreographed. Speculative theories abound. For instance, it has been proposed that flammable vegetation initially evolved in response to geological periods when atmospheric concentrations of O_2 were higher than at present (Robinson 1989). The diversification and geographic expansion of herbaceous species, and particularly grasses, since the mid Tertiary appears to have been stimulated by drier, colder global climates and possibly the physiological constraints of declining atmospheric CO_2 (Robinson 1994). This may have contributed to the evolution of flammable vegetation. Bond and van Wilgen (1996) note that 'grass fuels create a distinctive fire regime that must have acted as a powerful selective filter, permitting access for only a small subset of woody plants from ancestral forests'. Was the original stimulus for the evolution of *Eucalyptus* and allied species the ability to coexist with grass? In this context, it is interesting to note that the primary difference between *Allosyncarpia ternata* forests and surrounding *Eucalyptus* savannas is the degree of canopy closure. Both *Eucalyptus* and *A. ternata* have evergreen, sclerophyll foliage and have comparable drought tolerance. *A. ternata* canopies are closed because of the high density of horizontally arranged leaves, while *Euca-*

lyptus canopies are open due to the low density of pendulous leaves. *A. ternata* canopies intercept so much light that the development of a continuous grass layer is prevented, in marked contrast to *Eucalyptus* canopies (Bowman 1991*a*). Unlike *Eucalyptus* savanna, *A. ternata* is incapable of surviving recurrent fires typical of tropical savanna vegetation.

Clearly, far more research is required to explain the evolution of flammability. The flammability characteristics of genera with both rainforest and 'interspersive' non-rainforest species may provide insights into this evolutionary question (Table 12.1). One profitable approach may be to overlay characters that influence recovery and spread of fires on phylogenetic trees, particularly if phylogenies are derived from molecular data (Bond and van Wilgen 1996).

Conclusion

In the concluding chapter of their book *Fire and Plants*, Bond and van Wilgen (1996) asserted that 'the fire/climate relationship is not determined by climate alone. The fires that occur within a given climate depend upon the evolution or migration of flammable floras. The persistence of flammable and non-flammable formations under the same climate show that alternative solutions have evolved for the same physical circumstances'. This argument neatly encapsulates my belief that Australian rainforest and non-rainforest vegetation ultimately differ only by the degree which the constituent species are able to withstand recurrent fires. Once flammable vegetation had evolved in Australia, a positive-feedback cycle was established that resulted in the spread of 'pyrophytic' species and the corresponding contraction of 'pyrophobic' species to fire-resistant refugia (Ash 1988).

It is difficult to know what the trigger was for the evolution of flammable vegetation, although it was undoubtedly spurred by mid-Tertiary aridity. Aboriginal colonists entered a continent that was in the terminal stages of a long biological transformation forced by the interactive effects of aridity and the evolution of flammable forests. It is unknown if the Aboriginal colonists brought with them the cultural practice of using fire as a land-management tool or if this was developed following their arrival. However, there can be no doubt that their skilful use of fire provided a new dynamic in 'the closing chapters in this age-long struggle between these two great elements of the Australian flora' (Petrie *et al.* 1929). Indeed, were it not for the presence of Aborigines at the height of the last ice-age, the combined effect of fire and aridity may have been the *coup de grâce* for many species that had barely survived previous glacial cycles. While these views are necessarily speculative, it is clear that the maintenance of diversity in modern Australian vegetation, including rainforests, requires that humans actively influence the distribution and frequency of fire. Letting nature take her course may see the ultimate dominance of fire-adapted plants at the expense of the ancestral forests. Australians must use fire thoughtfully to manage their forest inheritance.

13

The fire theory VI. Fire management and rainforest conservation

Although there is an inextricable link between fire management and rainforest conservation, there is much contention about which fire regimes are appropriate. It is certainly the case that 'one size does not fit all'. Even similar rainforest types do not necessarily have the same fire management requirements, because the history of land use can be of overarching importance. For example, in spite of the ecological similarity of Northern Territory and Queensland monsoon rainforests, Russell-Smith and Bowman (1992) and Fensham (1996) came to very different conclusions about the consequences of current fire regimes. Russell-Smith and Bowman found that about one-third of the 1220 rainforest sites surveyed in the Northern Territory had boundaries severely degraded by fire, and concluded that this was the result of the near-complete breakdown of traditional Aboriginal fire management. In contrast, Fensham found that less than 10% of the 358 monsoon rainforest patches he surveyed in inland north Queensland were degraded by wildfire, and concluded that fire is not a major management issue. In north Queensland, intensive pastoralism has reduced grass fuels to such low levels that wildfire has become an unusual occurrence. However, Fensham noted that the situation in Queensland would change if pastoralists began to use fire to control native and exotic woody 'weeds' that have established on overgrazed rangelands.

In addition to directly threatening the biological integrity of many Australian rainforests, weeds also have the capacity to substantially change fuel characteristics across rainforest boundaries. For example, in inland north Queensland, weeds such as the introduced pasture grass *Cenchrus ciliaris*, the pasture weed *Parthenium hysterophorus*, the shrub *Lantana camara* and the vine *Cryptostegia grandiflora* increase the flammability of rainforest boundaries. Such weeds can become rampant on cleared land abutting fragments of rainforest, along bull-dozed fence lines, vehicle and mineral exploration tracks, and areas disturbed by pigs and cattle. Weed infestations can act as 'fire paths' into rainforest during drought periods. In the Northern Territory, Russell-Smith and Bowman (1992) found that the weeds *Hyptis suaveolens*, *Senna obtusifolia*, *Senna occidentalis*, *Sida acuta* and *Sida cordifolia* were often associated with rainforests damaged by fires, floods, storms and feral animals such as water buffalo, cattle and pigs. The relationships between these factors and weeds is complex as will be illustrated by the following two case studies.

Effect of fire, weeds and feral animal damage

Rainforests can be degraded as a result of the synergistic effects of feral animal disturbance, weed invasion and subsequent intense fires in unnaturally heavy fuel loads. Fensham *et al.* (1994) provided a clear example of this involving feral pigs and the shrub lantana (*Lantana camara*) in the Forty Mile Scrub, a mon-soonal rainforest 165 km inland from Cairns in north Queensland. Lantana was introduced into Australia from South America in the middle of the nineteenth century and has subsequently become a serious environmental weed. Fensham *et al.* found that lantana could establish impenetrable stands of more than 5000 stems per hectare, and that its density was negatively correlated with rainforest canopy cover and pig rooting. They postulated the following sequence of events to explain these patterns. Pig rooting kills the canopy trees by destroying their root systems. The consequent combination of disturbed soils and open canopy favours the establishment and ultimately the rampant growth of lantana. Dense stands of lantana produce heavy fuel loads across rainforest boundaries and permit savanna fires to penetrate the rainforest and kill any rainforest trees that have survived the pig rooting. Open conditions after the fire permit even greater dominance by lantana, further increasing fuel loads and perpetuating the cycle of rainforest destruction. The argument of Fensham *et al.* is supported by field experiments that were designed to determine the factors, such as soil nutrients, seedbed characteristics and light levels, that allow lantana infestations to estab-lish in dry rainforest in northeastern New South Wales (Gentle and Duggin 1997; Duggin and Gentle 1998). These authors clearly demonstrated that increased light levels, either following fire or mechanical removal of foliage, was the overarching requirement for lantana to successfully invade rainforest.

The management implication of Fensham *et al.*'s study is that, in order to conserve the Forty Mile Scrub, weeds, pigs and fire need to be controlled. They believed this goal could be best achieved by integrated management of pigs and fire in the surrounding landscape. Direct control of lantana is more difficult because there are no known biological control agents, and chemical or mechanical control is only practical on a small scale and is particularly difficult in a remote area such as Forty Mile Scrub.

In the humid subtropics where the fire risk is much less than in the Forty Mile Scrub, opinion is divided over the need to control lantana. Destruction of lantana can trigger the establishment of rainforest pioneer species such as *Omalanthus*, *Macaranga* and *Trema* (Nagle 1991). However, some authors have suggested that rainforest seedlings can germinate, or existing rootstocks re-shoot within, and ultimately over-top and shade out, stands of lantana (Dunphy 1991; Floyd 1991). One positive side effect of dense unburnt stands of lantana is that rainforest regeneration would be protected from pig rooting (Fensham *et al.* 1994).

Effect of fire, weeds and storm damage

Between 1945 and 1991 there has been a 60% reduction in the area of rainforest in the landscapes surrounding Darwin, the capital of the Northern Territory (Panton 1993). Two-thirds of the rainforest lost was destroyed in the development of the city, and the remaining one-third by the combined effects of tropical storm damage, weed invasion and intense fires. Panton noted that exotic vines such as *Calopoponium mucunoides*, *Clitoria ternata*, *Ipomoea quamoclit*, *Macroptilium* spp. and *Passiflora foetida* can be sufficiently rampant to smother trees on rainforest boundaries that have been degraded by storms and fires. When desiccated, these vines act as 'ladders' to carry fire into rainforest canopies.

However, the most serious weed in the Darwin area is the west African perennial grass *Pennisetum polystachion* (mission grass). It produces fuel loads about four times greater than found in weed-free *Eucalyptus* savannas, dries much later in the dry season than the native grasses, and when dry it maintains an upright stature, features that permit fires of far greater intensity than are usual for *Eucalyptus* savannas. If stands of *P. polystachion* are burnt earlier in the dry season when other fuels are still moist, sufficient fuel remains for another fire later in the same dry season. In one case documented by Panton (1993), *P. polystachion* established in a rainforest patch in the Darwin suburb of Leanyer following damage by a cyclone. The subsequent fire in the storm debris was so intense it destroyed 80% of the rainforest patch. Like lantana in the Forty Mile Scrub, *P. polystachion* can create a cycle of intense fires that are inimical to the survival of rainforest.

Since publishing his study, Panton (pers. comm.) has commenced a programme to rehabilitate the rainforest at Leanyer. He has found that, given sufficient time, replanted rainforest trees form a closed canopy that suppresses *P. polystachion* and consequently reduces the threat of fires. A central part of his strategy is the reduction of *P. polystachion* fuels in areas abutting replanted rainforest. This has been achieved by mechanically harvesting the grass or by burning it at night when fires are of a low enough intensity to be readily controlled. Panton also noted that achievement of effective fire management ultimately requires coordination amongst land-management agencies and a meshing of their aims and objectives to avoid the development of *ad hoc* and counterproductive land and fire management practices.

The use of fire for the management and conservation of non-rainforest vegetation in rainforest areas

There are situations where the primary focus of fire management is not rainforest but the surrounding vegetation. This can result in the conflict of management objectives. Gullies containing rainforest are sometimes used as firebreaks to limit fires used to reduce fuel loads or to prepare seedbeds after logging (Adam 1992). Applied research is needed to enable managers to use fire without inadvertently degrading rainforest. For example, Marsden-Smedley and Catchpole (1995*a,b*) studied the fire behaviour of fire-prone sedgelands in southwest Tasmania, providing managers with the knowledge required to burn the sedgelands without threatening surrounding forests.

In some environments, European fire regimes have led to the expansion of rainforest. Harrington (1995) has described the invasion of tall *Eucalyptus grandis* forests by tropical rainforest. The *E. grandis* forests form ecotones less than one km wide between the rainforest and drier *Eucalyptus* forest on the western edge of the Atherton Tablelands in Queensland, where they are an important habitat for arboreal mammals such as the northern race (*reginae*) of the Yellow-bellied Glider (*Petaurus australis*). It is thought that the rainforest has expanded at the expense of the *E. grandis* forests because fires are now of low intensity following the reduction of fuel loads by cattle and the regular use of fire by pastoralists to regenerate grass, whereas *E. grandis* forests require occasional high-intensity fires to destroy the emerging rainforest understorey. This is a dilemma for land managers because high-intensity fires are very difficult to control. Furthermore, high-intensity fires also destroy the hollow eucalypts upon which the gliders depend for shelter, so that the requirements of the Yellow-bellied Glider can only be maintained with a mosaic of *E. grandis* forests of various fire ages.

Global climate change and associated elevated CO_2 levels (the 'fertiliser effect') may result in substantial changes in the structure and species composition of forests (Phillips and Gentry 1994; Phillips 1995; Sheil 1995). If this effect proves to be real then it is possible that rainforest may rapidly colonise surrounding *Eucalyptus* forests in some areas, further increasing the need for the use of fire to maintain habitat diversity.

Research to bolster management

The ecological basis for the use of fire in Australian landscape management was not widely appreciated until the 1970s (Luke and McArthur 1978; Gill *et al.* 1981). Much more research is required to understand the relationship between fire and rainforests. In environments where Aboriginal people maintain close links with the land, such as in parts of the monsoon tropics, there is an urgent need to involve them in collaborative research on fire ecology. It is particularly important that the knowledge of older generations of Aboriginal people is documented before it is lost forever. Some of these people experienced an essentially traditional lifestyle before the massive disruptions that accompanied European colonisation in the mid twentieth century. A fine example of such research is the study of Russell-Smith *et al.* (1997*a*). In landscapes that have been depopulated of Aboriginal people and which have suffered drastic environment- al transformations following European colonisation, pure and applied research is required to determine the past and present effects of fire regimes on rainforest boundaries. Much more research is required to understand the role of fire in creating appropriate vegetation mosaics for wildlife (Johnson 1980; Vernes *et al.* 1995).

One of the most basic levels of information required to manage the impact of fire on rainforest is a method to identify rainforests most at risk from fire damage. This 'rainforest triage' is urgent because land managers must balance the need to manage entire landscapes against the need to manage individual rainforest patches. Russell-Smith and Bowman (1992) noted that fire manage- ment in Northern Territory landscapes typically consists of broad scale 'control- led burning' early in the dry season, an objective usually accomplished by dropping incendiaries during traverses by helicopters or fixed-wing aircraft. The method is seen as a cost-effective and practical way of ensuring that widespread high-intensity fires do not sweep across the savanna later in the dry season. However, this style of fire management often leaves tracts of unburnt savanna abutting rainforest patches, rendering rainforests susceptible to fire damage later in the dry season. Russell-Smith and Bowman argued that the 'inescapable conclusion for managers is that biologically significant patches [of rainforest] require individual attention'. However, it has been estimated that there are some 15 000 patches of monsoon rainforest in the Northern Territory. Unlike Aboriginal people who in the past led a nomadic hunter–gatherer lifestyle, it is impossible for today's land managers to individually burn around rainforest patches. Further, the archipelago of rainforest habitats requires management at a landscape scale because the long-term survival of isolated and small populations of rainforest plants appears to depend upon genetic exchange between patches (Price *et al.* 1998). Birds and bats are important long-distance vectors of both pollen and seeds, and the long-term future of these vertebrates is also linked to the fate of the diaspora of rainforest habitats.

On-ground monitoring using permanent plots is one way to assess the condition of rainforest. However, in northern Australia this is impractical because of the vastness of the landscapes. Remote sensing offers an alternative approach. Although satellite imagery is often too coarse, digitised aerial photographs that have been spatially rectified can be used to precisely monitor changes in the extent of rainforest vegetation. Statistical relationships between environmental data and the direction and magnitude of change to individual rainforest boundaries could be used to identify rainforest at risk.

Conclusion

Conservation of the Australian rainforest inheritance depends on fire management, which must be crafted to suit both the individual rainforest patch and the ecology of the surrounding landscapes. Because resources are limited, there is an urgent need to identify rainforest patches that require immediate attention. Meeting rainforest conservation objectives will require much creative research and good communication and coordination amongst land managers.

14

Summary

Rainforests in Australia occur as scattered islands within vast tracts of the quintessential Australian vegetation dominated by *Eucalyptus* and *Acacia*. Rainforest 'islands' are found from Tasmania in the temperate zone to the monsoon tropics in the north across wide gradients of rainfall, altitude and temperature. In this book, I argue that their patchy distribution is the consequence of tens of millions of years of fires started by lightning and other natural causes. Fire became a feature of the Australian landscape as the continent became progressively drier. This occurred because of the dominance of the subtropical high pressure system beginning in the mid Tertiary when the continent drifted into the mid latitudes, the thermal gradient between the equator and the South Pole intensified, and the Antarctic ice sheet formed. An increase in the frequency of wildfires triggered a major diversification of the fire-tolerant biota. In plants, there was selection for new traits such as serotiny, and the refinement of pre-existing traits such as the lignotuber. Some fire-tolerant vegetation became fire-adapted and ultimately fire-promoting, sealing the fate of most rainforest vegetation. This evolutionary divergence is responsible for the fundamental and often sharp ecological and floristic dichotomy of the current Australian biota into rainforest and non-rainforest types. Thus the evolution of the current Australian flora was forged in a fiery environment in which the ancestral rainforests become scattered and burnt.

I reject the hypothesis that the shift in dominance from rainforest to non-rainforest vegetation and the diversification of fire-tolerant flora and fauna was the result of fires lit by the Aborigines following their arrival in Australia during the last glacial. The ice-age Australians colonised a continent that was already ablaze, and the triumph of these colonists was in harnessing landscape fire to their own ends. In altering the frequency and intensity of fires, they undoubtedly influenced the distribution of vegetation types. Grassland and *Eucalyptus* forests within tracts of rainforest are a legacy of the use of fire by Aboriginal people. However, the skilful management of fire by the Australian Aborigines may have served to maintain much of the rainforest heritage.

In this book, I have evaluated a range of evidence and considered alternative theories that have been advanced to explain differences in the nature and distribution of the rainforest and non-rainforest biota. This critical analysis is frustrated by confusion about the definition of rainforest, which goes back to the roots of Australian botanical scholarship. The original concept of rainforest, formulated by Schimper at the end of the nineteenth century, explicitly included tall *Eucalyptus* forest in high rainfall climates, whereas 'sclerophyll forests', a term also coined by Schimper, included *Eucalyptus* forests from drier climates. Australian ecologists after Schimper have generally chosen to refer to tall *Eucalyptus* forests as 'wet sclerophyll forests' and to contrast them with rainforests. They did so because of the remarkable morphological uniformity of *Eucalyptus* throughout its climatic range, and the starkness of the contrast with 'true' rainforest trees. This reasoning followed, wittingly or unwittingly, Hooker's nineteenth century dichotomization of the Australian flora, in which he noted that the feature distinguishing the 'sclerophyll' and rainforest floras was the bizarre morphology of the former and not any real phylogenetic uniqueness. The naming of Botany Bay, widely regarded as the birthplace of the modern Australian nation, attests to the considerable impact the 'strange' Australian flora had on the early botanists. Hooker acknowledged the existence of isolated pockets of 'normal' vegetation in Australia and assumed that these rainforests, to use modern parlance, were invasive.

I devoted a chapter each to the meaning of the terms 'rainforest' and 'sclerophyll'. I have demonstrated that there is no simple conceptual demarcation of Australian rainforest. Although it is generally accepted that rainforests have closed canopies that contrast with the open canopies of non-rainforest vegetation, definitions of rainforest often also emphasise various other biological characters including life forms, structure, the biogeographic affinities of the dominant taxa, and regeneration strategies. A recurring feature is the incorporation of ecological theories into the definitions, though these are sometimes concealed like a Trojan horse. I argued that the definitional problem cannot be separated from issues relating to the ecological and historical biogeography of Australian forests.

The remainder of the book was devoted to critically evaluating the ecological hypotheses that have been proffered to explain the distribution of Australian forests, especially rainforests. In summary, these are:

(i) rainforest can only occur on soils of high fertility. A special case of this theory is that the critical nutrient is soil phosphorus;

(ii) rainforest can only occur on perennially moist soils because the vegetation must maintain continuous transpiration;

(iii) rainforest trees require sheltered microclimates for growth and especially germination, and are therefore excluded from the exposed microclimates of adjacent non-rainforest vegetation; and

(iv) rainforest can only occur in topographic settings protected from landscape fires.

Some authors have emphasised a few or all the possible interactions between these. I have explored the hypotheses in detail, and shown that the evidence does not support the generalisation that soil fertility, microclimate, or moisture supply limits rainforest vegetation. It should be noted that this conclusion is reached by using the various and disparate definitions of rainforest employed by the authors of the research that I reviewed. It is possible however, to substantiate the generalisation that fire is a critical factor in limiting rainforest in all forest environments throughout Australia. Even this generalisation requires qualification, as rainforest trees are shown to have a surprising capacity to recover vegetatively following fire damage, and some rainforests occur in areas with no fire protection. The frequency of fire in the landscape is of critical importance in controlling the distribution of rainforest. Frequent fires prevent the maturation of seedlings or re-sprouts, and ultimately eliminate rainforest from a landscape except where the topography creates fire refugia.

Historical theories that attempt to explain the fragmentation of rainforest and the evolution of non-rainforest vegetation focus either on climate change through the Cainozoic Era or the increased occurrence of fire following human colonisation in the Pleistocene Epoch. There is some evidence that non-rainforest vegetation can be legitimately described as pyrophytic in that it has not only evolved traits to enable rapid recovery following fire, but has also evolved features that increase the probability of fires. Conversely, rainforest vegetation can be legitimately described as pyrophobic as it is less able to recover from fire and has characteristics that render it less susceptible to fire. Natural selection for traits permitting survival and the promotion of wildfire appears to have been driven by aridification of the Australian continent since the mid Tertiary and fires ignited by natural causes.

It was in this evolutionary context that humans colonised Australia. There is no unambiguous evidence with which to determine the impact of the Pleistocene colonists on rainforest. This is attributable in no small part to uncertainty about the timing of colonisation. It is my view that fire management by the ice-age Australians was critical in conserving some fire-sensitive vegetation during the height of the Pleistocene aridity 20000 years ago, but given the dearth of evidence this view is necessarily speculative. Nonetheless, there is some evidence that Aboriginal landscape burning influenced the pattern and rate of expansion of rainforest when the late-Pleistocene aridity gave way to the warmer and wetter climates of the Holocene. There is clear evidence that, at the time of European colonisation, Aborigines carefully managed landscapes with fire and the extent of rainforest was influenced by their burning practices. Following European colonisation, the cessation of Aboriginal landscape burning has led to the expansion of rainforest in some places and its contraction in others.

Management of the Australian rainforest estate demands that fire is used carefully and deliberately to maintain rainforest boundaries. If we let nature take her course, we may see the ultimate dominance of fire-adapted plants at the expense of the ancestral and less fire-adapted rainforests in all but the most humid climates and best-protected fire refugia. Land clearance, rampant weed infestation, disturbance by feral animals and settlements in or close to bushland confound simple solutions. The conflicting objectives of different interest groups including governmental agencies are serious impediments to constructive management of rainforests and fire. There is an urgent need for creative research to facilitate and prioritise fire management to ensure the long-term conservation of rainforest and surrounding habitat mosaics.

References

Adam, P. (1992). *Australian rainforests.* (Clarendon Press: Oxford.)

Adam, P., Stricker, P. and Anderson, D.J. (1989). Species-richness and soil phosphorus in plant communities in coastal New South Wales. *Australian Journal of Ecology* 14, 189–198.

Adamson, R.S. and Osborn, T.G.B. (1924). The ecology of the *Eucalyptus* forests of the Mount Lofty Ranges (Adelaide district), South Australia. *Royal Society of South Australia* 48, 87–143.

Ahern, C.R. and Macnish, S.E. (1983). Comparative study of phosphorus and potassium levels of basaltic soils associated with scrub and forest communities on the Darling Downs. *Australian Journal of Soil Research* 21, 527–538.

Allen, J. (1994). Radiocarbon determinations, luminescence dating and Australian archaeology. *Antiquity* 68, 339–349.

Allen, J. and Holdaway, S. (1995). The contamination of Pleistocene radiocarbon determinations in Australia. *Antiquity* 69, 101–112.

Andersen, A.N. (1991). Response of ground-foraging ant communities to three experimental fire regimes in a savanna forest of tropical Australia. *Biotropica* 23, 575–585.

Anderson, A. (1994). Comment on J. Peter White's paper "Site 820 and the evidence for early occupation in Australia". *Quaternary Australasia* 12, 30–31.

Andrews, E.C. (1916). The geological history of the Australian flowering plants. *American Journal of Science* 42, 171–232.

Anonymous (1986). *Atlas of Australian Resources. Volume 4: Climate.* (Division of National Mapping: Canberra.)

Anonymous (1988). *Atlas of Australian Resources. Volume 5: Geology and Minerals.* (Australian Surveying and Land Information Group: Canberra.)

Anonymous (1990). *Atlas of Australian Resources. Volume 6: Vegetation.* (Australian Surveying and Land Information Group: Canberra.)

Anonymous (1991). *The distribution of Huon Pine* Lagarostrobos franklinii, *King*

Billy Pine Athrotaxis selaginoides *and Deciduous Beech* Nothofagus gunnii *in Tasmania.* (Forestry Commission of Tasmania: Hobart.)

Archer, M., Godthelp, H., Hand, S.J. and Megirian, D. (1989). Fossil mammals of Riversleigh, northwestern Queensland: Preliminary overview of biostratigraphy, correlation and environmental change. *Australian Zoologist* 25, 29–65.

Archer, M., Hand, S.J. and Godthelp, H. (1994). Patterns in the history of Australia's mammals and inferences about palaeohabitats. In *History of the Australian Vegetation: Cretaceous to Recent* (ed. R.S. Hill) pp. 80–103. (Cambridge University Press: Cambridge.)

Ash, J. (1988). The location and stability of rainforest boundaries in north-eastern Queensland, Australia. *Journal of Biogeography* 15, 619–630.

Ashton, D.H. (1975). Studies of litter in *Eucalyptus regnans* forests. *Australian Journal of Botany* 23, 413–433.

Ashton, D.H. (1976a). Phosphorus in forest ecosystems at Beenak, Victoria. *Journal of Ecology* 64, 171–186.

Ashton, D.H. (1976b). Studies on the Mycorrhizae of *Eucalyptus regnans* F. Muell. *Australian Journal of Botany* 24, 723–741.

Ashton, D.H. (1981a). Fire in tall open-forest (wet sclerophyll forests). In *Fire and the Australian Biota* (eds. A.M. Gill, R.H. Groves and I.R. Noble) pp. 339–366. (Australian Academy of Sciences: Canberra.)

Ashton, D.H. (1981b). Tall open-forests. In *Australian Vegetation* (ed. R.H. Groves) pp. 121–151. (Cambridge University Press: Cambridge.)

Ashton, D.H. and Frankenberg, J. (1976). Ecological studies of *Acmena smithii* (Poir.) Merrill & Perry with special reference to Wilson's Promontory. *Australian Journal of Botany* 24, 453–487.

Ashton, D.H. and Turner, J.S. (1979). Studies on the light compensation point of *Eucalyptus regnans* F. Muell. *Australian Journal of Botany* 27, 589–607.

Ashton, D.H. and Willis, E.J. (1982). Antagonisms in the regeneration of *Eucalyptus regnans* in the mature forest. In *The Plant Community as a Working Mechanism* (ed. E.I. Newman) pp. 113–128. (Blackwell Scientific Publications: Oxford.)

Attiwill, P.M. (1968). The loss of elements from decomposing litter. *Ecology* 49, 142–145.

Attiwill, P.M. and Clayton-Greene, K.A. (1984). Studies of gas exchange and development in a subhumid woodland. *Journal of Ecology* 72, 285–294.

Attiwill, P.M., Guthrie, H.B. and Leuning, R. (1978). Nutrient cycling in a *Eucalyptus obliqua* (L'Herit.) forest. I Litter production and nutrient return. *Australian Journal of Botany* 26, 79–91.

Austin, M.P., Pausas, J.G. and Noble, I.R. (1997). Modelling environmental and temporal niches of eucalypts. In *Eucalypt ecology: individuals to ecosystems* (eds. J.E. Williams and J.C.Z. Woinarski) pp. 129–150. (Cambridge University Press: Cambridge.)

Barber, H.N. and Jackson, W.D. (1957). Natural selection in action in *Eucalyptus*. *Nature* 179, 1267–1269.

Barker, M.J. (1991). *The effect of fire on west coast lowland rainforest.* (Tasmanian National Rainforest Conservation Program Report No. 7: Canberra.)

Barker, P.C.J. (1991). *Podocarpus lawrencei* (Hook. f.): Population structure and fire history at Goonmirk Rocks, Victoria. *Australian Journal of Ecology* 16, 149–158.

Barlow, B.A. and Hyland, B.P.M. (1988). The origins of the flora of Australia's wet tropics. *Proceedings of the Ecological Society of Australia* 15, 1–17.

Barrett, D.J. and Ash, J.E. (1992). Growth and carbon partitioning in rainforest and eucalypt forest species of south coastal New South Wales, Australia. *Australian Journal of Botany* 40, 13–25.

Barrett, D.J., Hatton, T.J., Ash, J.E. and Ball, M.C. (1996). Transpiration by trees from contrasting forest types. *Australian Journal of Botany* 44, 249–263.

Barrow, P., Duff, G., Liddle, D. and Russell-Smith, J. (1993). Threats to monsoon rainforest habitat in northern Australia: the case of *Ptychosperma bleeseri* Burret (Arecaceae). *Australian Journal of Ecology* 18, 463–471.

Baur, G.N. (1957). Nature and distribution of rain-forests in New South Wales. *Australian Journal of Botany* 5, 190–233.

Baur, G.N. (1965). *Forest types in New South Wales.* (Forestry Commission of New South Wales Research Note 17: Sydney.)

Baur, G.N. (1968). *The ecological basis of rainforest management.* (Government Printer: Sydney.)

Baverstock, P.R., Richardson, B.J., Birrell, J. and Krieg, M. (1989). Albumin immunologic relationships of the Macropodidae (Marsupialia). *Systematic Zoology* 38, 38–50.

Beadle, N.C.W. (1953). The edaphic factor in plant ecology with a special note on soil phosphates. *Ecology* 34, 426–428.

Beadle, N.C.W. (1954). Soil phosphate and the delimitation of plant communities in eastern Australia. *Ecology* 35, 370–375.

Beadle, N.C.W. (1962*a*). Soil phosphate and the delimitation of plant communities in eastern Australia. II. *Ecology* 43, 281–288.

Beadle, N.C.W. (1962*b*). An alternative hypothesis to account for the generally low phosphate content of Australian soils. *Australian Journal of Agricultural Research* 13, 434–442.

Beadle, N.C.W. (1966). Soil phosphate and its role in moulding segments of the Australian flora and vegetation, with special reference to xeromorphy and sclerophylly. *Ecology* 47, 992–1007.

Beadle, N.C.W. (1968). Some aspects of the ecology and physiology of Australian xeromorphic plants. *Australian Journal of Science* 30, 348–355.

Beadle, N.C.W. (1981). Origins of the Australian angiosperm flora. In *Ecological biogeography of Australia* (ed. A. Keast) pp. 407–426. (Dr W. Junk: The Hague.)

Beadle, N.C.W. and Costin, A.B. (1952). Ecological classification and nomenclature. *Proceedings of the Linnean Society of New South Wales* 77, 61–82.

Beard, J.S. (1955). The classification of tropical American vegetation-types. *Ecology* **36**, 89–100.

Beard, J.S. (1976). The monsoon forests of the Admirality Gulf, Western Australia. *Vegetatio* **31**, 177–192.

Beard, J.S. (1990). Temperate forests of the southern hemisphere. *Vegetatio* **89**, 7–10.

Bjorkman, O. and Ludlow, M.M. (1972). Characterization of the light climate on the floor of a Queensland rainforest. *Carnegie Institution of Washington Year Book* **71**, 85–94.

Blake, S.T. (1938). The plant communities of western Queensland and their relationships, with special reference to the grazing industry. *Proceedings of the Royal Society of Queensland* **49**, 156–204.

Bond, W.J. and Midgely, J.J. (1995). Kill thy neighbour: an individualistic argument for the evolution of flammability. *Oikos* **73**, 79–85.

Bond, W.J. and van Wilgen, B.W. (1996). *Fire and plants.* (Chapman and Hall: London.)

Boon, S. and Dodson, J.R. (1992). Environmental response to land use at Lake Curlip, East Gippsland, Victoria. *Australian Geographical Studies* **30**, 206–221.

Bowdler, S. (1983). Rainforest: colonised or coloniser? *Australian Archaeology* **17**, 59–66.

Bowdler, S. (1991). Some sort of dates at Malakunanja II: a reply to Roberts *et al. Australian Archaeology* **32**, 50–51.

Bowler, J.M. (1982). Aridity in the late Tertiary and Quaternary of Australia. In *Evolution of the flora and fauna of arid Australia* (eds. W.R. Barker and P.J.M. Greenslade) pp. 35–45. (Peacock Publications: Adelaide.)

Bowman, D.M.J.S. (1986). Stand characteristics, understorey associates and environmental correlates of *Eucalyptus tetrodonta* F. Muell. forests on Gunn Point, northern Australia. *Vegetatio* **65**, 105–113.

Bowman, D.M.J.S. (1988). Stability amid turmoil?: towards an ecology of north Australian eucalypt forests. *Proceedings of the Ecological Society of Australia* **15**, 149–158.

Bowman, D.M.J.S. (1991*a*). Environmental determinants of *Allosyncarpia ternata* forests that are endemic to western Arnhem Land, northern Australia. *Australian Journal of Botany* **39**, 575–589.

Bowman, D.M.J.S. (1991*b*). Recovery of some northern Australian monsoon forest tree species following fire. *Proceedings of the Royal Society of Queensland* **101**, 21–25.

Bowman, D.M.J.S. (1992*a*). Monsoon forests in north-western Australia. II. Forest–savanna transitions. *Australian Journal of Botany* **40**, 89–102.

Bowman, D.M.J.S. (1992*b*). Towards a definition of Australian rainforest types: The Northern Territory Experience. In *Victoria's rainforests: perspectives on definition, classification and management* (eds. P. Gell and D. Mercer) pp. 163–165. (Monash Publications in Geography No. 41: Melbourne.)

Bowman, D.M.J.S. (1992*c*). Evidence for gradual retreat of dry monsoon forests under a regime of Aboriginal burning, Karslake Peninsula, Melville Island, northern Australia. *Proceedings of the Royal Society of Queensland* 102, 25–30.

Bowman, D.M.J.S. (1993). Establishment of two dry monsoon forest tree species on a fire-protected monsoon forest-savanna boundary, Cobourg Peninsula, northern Australia. *Australian Journal of Ecology* 18, 235–237.

Bowman, D.M.J.S. (1994). Preliminary observations on the mortality of *Allosyncarpia ternata* stems on the Arnhem Land Plateau, northern Australia. *Australian Forestry* 57, 62–64.

Bowman, D.M.J.S. (1995). Two examples of the role of ecological biogeography in Australian prehistory: The fire ecology of *Callitris intratropica*, and the spatial pattern of stone tools in the Northern Territory. *Australian Archaeology* 41, 8–11.

Bowman, D.M.J.S. (1998). Tansley Review 101: The impact of Aboriginal landscape burning on the Australian biota. *New Phytologist* 140, 385–410.

Bowman, D.M.J.S. and Brown, M.J. (1986). Bushfires in Tasmania: a botanical approach to anthropological questions. *Archaeology in Oceania* 21, 166–171.

Bowman, D.M.J.S. and Connors, G.T. (1996). Does low temperature cause the dominance of *Acacia* on the central Australian mountains? Evidence from a latitudinal gradient from 11° to 26° South in the Northern Territory, Australia. *Journal of Biogeography* 23, 245–256.

Bowman, D.M.J.S and Dunlop, C.R. (1986). Vegetation pattern and environmental correlates in coastal forests of the Australian monsoon tropics. *Vegetatio* 65, 99–104.

Bowman, D.M.J.S. and Fensham, R.J. (1991). Response of a monsoon forest–savanna boundary to fire protection, Weipa, northern Australia. *Australian Journal of Ecology* 16, 111–118.

Bowman, D.M.J.S. and Fensham, R.J. (1995). Growth of *Eucalyptus tetrodonta* seedlings on savanna and monsoon rainforest soils in the Australian monsoon tropics. *Australian Forestry* 58, 46–47.

Bowman, D.M.J.S. and Harris, S. (1995). Conifers of Australia's dry forests and open woodlands. In *Ecology of the southern conifers* (eds. N.J. Enright and R.S. Hill) pp. 252–270. (University of Melbourne Press: Melbourne.)

Bowman, D.M.J.S. and Jackson, W.D. (1981). Vegetation succession in southwest Tasmania. *Search* 12, 358–362.

Bowman, D.M.J.S. and Latz, P.K. (1993). Ecology of *Callitris glaucophylla* (Cupressaceae) on the MacDonnell Ranges, central Australia. *Australian Journal of Botany* 41, 217–225.

Bowman, D.M.J.S., Latz, P.K. and Panton, W.J. (1994a). Pattern and change in an *Acacia aneura* shrubland and *Triodia* hummock grassland mosaic on rolling hills in central Australia. *Australian Journal of Botany* 43, 25–37.

Bowman, D.M.J.S., Maclean, A.R., and Crowden, R.K. (1986). Vegetation–soil relationships in the lowlands of south-west Tasmania. *Australian Journal of Ecology* 11, 141–153.

Bowman, D.M.J.S. and McDonough, L. (1991). Tree species distribution across a seasonally flooded elevation gradient in the Australian monsoon tropics. *Journal of Biogeography* 18, 203–212.

Bowman, D.M.J.S. and Minchin, P.R. (1987). Environmental relationships of woody vegetation patterns in the Australian monsoon tropics. *Australian Journal of Botany* 35, 151–169.

Bowman, D.M.J.S. and Panton, W.J. (1993a). Decline of *Callitris intratropica* R.T. Baker & H.G. Smith in the Northern Territory: implications for pre- and post-European colonization fire regimes. *Journal of Biogeography* 20, 373–381.

Bowman, D.M.J.S. and Panton, W.J. (1993b). Factors that control monsoon-rainforest seedling establishment and growth in north Australian *Eucalyptus* savanna. *Journal of Ecology* 81, 297–304.

Bowman, D.M.J.S. and Panton, W.J. (1993c). Differences in the stand structure of *Eucalyptus tetrodonta* forests between Elcho Island and Gunn Point, northern Australia. *Australian Journal of Botany* 41, 211–215.

Bowman, D.M.J.S. and Panton, W.J. (1994). Fire and cyclone damage to woody vegetation on the north coast of the Northern Territory, Australia. *Australian Geographer* 25, 32–35.

Bowman, D.M.J.S. and Panton, W.J. (1995). Munmarlary revisited: response of a north Australian *Eucalyptus tetrodonta* savanna protected from fire for 20 years. *Australian Journal of Ecology* 20, 526–531.

Bowman, D.M.J.S, Panton, W.J. and Head, J. (1999a). Abandoned Orange-footed Scrubfowl (*Megapodius reinwardt*) nests and coastal rainforest boundary dynamics during the late Holocene in monsoonal Australia. *Quaternary International* 59, 27–38.

Bowman, D.M.J.S., Panton, W.J. and McDonough, L. (1990). Dynamics of forest clumps on chenier plains, Cobourg Peninsula, Northern Territory. *Australian Journal of Botany* 38, 593–601.

Bowman, D.M.J.S. and Rainey, I. (1995). Tropical tree stand structures on a seasonally flooded elevation gradient in northern Australia. *Australian Geographer* 27, 31–37.

Bowman, D.M.J.S. and Wightman, G.M. (1985). Small scale vegetation pattern associated with a deeply incised gully, Gunn Point, northern Australia. *Proceedings of the Royal Society of Queensland* 96, 63–73.

Bowman, D.M.J.S. and Wilson, B.A. (1988). Fuel characteristics of coastal monsoon forests, Northern Territory, Australia. *Journal of Biogeography* 15, 807–817.

Bowman, D.M.J.S., Wilson, B.A. and Dunlop, C.R. (1988a). Preliminary biogeographic analysis of the Northern Territory vascular flora. *Australian Journal of Botany* 36, 503–517.

Bowman, D.M.J.S., Wilson, B.A. and Fensham, R.J. (1999b). Relative drought tolerance of evergreen-rainforest and evergreen-savanna species in a long unburnt *Eucalyptus* savanna, north Queensland. *Proceedings of the Royal Society of Queensland* 108, 27–31.

Bowman, D.M.J.S., Wilson, B.A. and Hooper, R.J. (1988b). Response of *Eucalyptus* forest and woodland to four fire regimes at Munmarlary, Northern Territory, Australia. *Journal of Ecology* 76, 215–232.

Bowman, D.M.J.S., Wilson, B.A. and McDonough, L. (1991*a*). Monsoon forests in northwestern Australia I. Vegetation classification and the environmental control of tree species. *Journal of Biogeography* 18, 679–686.

Bowman, D.M.J.S., Wilson, B.A. and Woinarski, J.C.Z. (1991*b*). Floristic and phenological variation in a northern Australian rocky *Eucalyptus* savanna. *Proceedings of the Royal Society of Queensland* 101, 79–90.

Bowman, D.M.J.S. and Woinarski, J.C.Z. (1994). Biogeography of Australian monsoon rainforest mammals: implications for the conservation of rainforest mammals. *Pacific Conservation Biology* 1, 98–106.

Bowman, D.M.J.S., Woinarski, J.C.Z. and Menkhorst, K.A. (1993). Environmental correlates of tree species diversity in Stage III of Kakadu National Park, northern Australia. *Australian Journal of Botany* 41, 649–660.

Bowman, D.M.J.S., Woinarski, J.C.Z. and Russell-Smith, J. (1994*b*). Environmental relationships of Orange-footed Scrubfowl *Megapodius reinwardt* nests in the Northern Territory. *Emu* 94, 181–185.

Bradstock, R.A., Bedward, M., Scott, J. and Keith, D.A. (1996). Simulation of the effect of spatial and temporal variation in fire regimes on the population viability of a *Banksia* species. *Conservation Biology* 10, 776–784.

Braithwaite, L.W., Dudzinski, M.L. and Turner, J. (1983). Studies on the arboreal marsupial fauna of eucalypt forest being harvested for woodpulp at Eden, N.S.W. II. Relationship between the fauna density, richness and diversity, and measured variables of the habitat. *Australian Wildlife Research* 10, 231–247.

Braithwaite, R.W. (1991). Aboriginal fire regimes of monsoonal Australia in the 19th century. *Search* 22, 247–249.

Braithwaite, R.W., Dudzinski, M.L., Ridpath, M.G. and Parker, B.S. (1984). The impact of water buffalo on the monsoon forest ecosystem in Kakadu National Park. *Australian Journal of Ecology* 9, 309–322.

Brough, P., McLuckie, J. and Petrie, A.H.K. (1924). An ecological study of the flora of Mount Wilson. *Proceedings of the Linnean Society of New South Wales* 49, 475–498.

Brown, M.J. (1988). *The distribution and conservation of King Billy Pine*. (Forestry Commission of Tasmania.)

Brown, M.J. (1996). Benign neglect and active management in Tasmania's forests: a dynamic balance or ecological collapse? *Forest Ecology and Management* 85, 279–289.

Brown, M.J. and Podger, F. (1982*a*). On the apparent anomaly between observed and predicted percentages of vegetation types in south-west Tasmania. *Australian Journal of Ecology* 7, 203–205.

Brown, M.J. and Podger, F.D. (1982*b*). Floristics and fire regimes of a vegetation sequence from sedgeland-heath to rainforest at Bathurst Harbour, Tasmania. *Australian Journal of Botany* 30, 659–676.

Brown, M.J., Ratkowsky, D.A. and Minchin, P.R. (1984). A comparison of detrended correspondence analysis and principal co-ordinates analysis using four sets of Tasmanian vegetation data. *Australian Journal of Ecology* **9**, 273–279.

Brown, N. (1998). Out of control: fires and forestry in Indonesia. *Trends in Ecology and Evolution* **13**, 41.

Buckley, R. (1984). The role of fire: response to Synder. *Oikos* **43**, 405–406.

Burges, A. and Johnston, R.D. (1953). The structure of a New South Wales subtropical rain forest. *Journal of Ecology* **41**, 72–83.

Busby, J.R. (1986). A biogeoclimatic analysis of *Nothofagus cunninghamii* (Hook.) Oerst. in southeastern Australia. *Australian Journal of Ecology* **11**, 1–7.

Cameron, D. (1992). A portrait of Victoria's rainforests: distribution, diversity and definition. In *Victoria's rainforests: perspectives on definition, classification and management* (eds. P. Gell and D. Mercer) pp. 13–50. (Monash Publications in Geography No. 41: Melbourne.)

Cameron, R.J. (1970). Light intensity and the growth of *Eucalyptus* seedlings. I. Ontogenetic variation in *E. fastigata*. *Australian Journal of Botany* **18**, 29–43.

Carnahan, J.A. (1981). Mapping at a continental level. In *Vegetation Classification in Australia* (eds. A.N. Gillison and D.J. Anderson) pp. 107–113. (Commonwealth Scientific and Industrial Research Organisation and Australian National University: Canberra.)

Chandler, G.E. and Lamb, D. (1986). The role of plant nutrition in the functioning of plant communities in northern Australia—some hypotheses. In *Tropical plant communities: their resilience, functioning and management in northern Australia* (eds. H.T. Clifford and R.L. Specht) pp. 53–67. (Department of Botany, University of Queensland: Brisbane.)

Chappell, J. (1991). Late Quaternary environmental changes in eastern and central Australia, and their climatic interpretation. *Quaternary Science Reviews* **10**, 377–390.

Chappell, J., Head, J. and Magee, J. (1996). Beyond the radiocarbon limit in Australian archaeology and Quaternary research. *Antiquity* **70**, 543–552.

Chen Y. (1988). Early Holocene population expansion of some rainforest trees at Lake Barrine basin, Queensland. *Australian Journal of Ecology* **13**, 225–233.

Chippendale, G.M. (1963). The relic nature of some central Australian plants. *Transactions of the Royal Society of South Australia* **86**, 31–34.

Chippendale, G.M and Wolf, L. (1981). *The natural distribution of Eucalyptus in Australia*. (Australian National Parks and Wildlife Service Special Publication No. 6: Canberra.)

Choong, M.F., Lucas, P.W., Ong, J.S.Y., Pereira, B., Tan, H.T.W. and Turner, I.M. (1992). Leaf fracture toughness and sclerophylly: their correlations and ecological implications. *New Phytologist* **121**, 597–610.

Choquenot, D. and Bowman, D.M.J.S. (1998). Marsupial megafauna, Aborigines and the overkill hypothesis: application of predator-prey models to the question of Pleistocene extinction in Australia. *Global Ecology and Biogeography Letters* **7**, 167–180.

Christidis, L. and Schodde, R. (1993). Relationships and radiations in the meliphagine honeyeaters, *Meliphaga*, *Lichenostomus* and *Xanthotis* (Aves: Meliphagidae): Protein evidence and its integration with morphology and ecogeography. *Australian Journal of Zoology* 41, 293–316.

Churchill, S.K. (1991). Distribution, abundance and roost selection of the Orange Horseshoe-bat, *Rhinonycteris aurantius*, a tropical cave-dweller. *Wildlife Research* 18, 343–353.

Clark, R.L. (1983). Pollen and charcoal evidence for the effects of Aboriginal burning on the vegetation of Australia. *Archaeology in Oceania* 18, 32–37.

Clayton-Greene, K.A. (1983). The tissue water relationships of *Callitris columellaris*, *Eucalyptus melliodora* and *Eucalyptus microcarpa* investigated using the pressure-volume technique. *Oecologia* 57, 368–373.

Clayton-Greene, K.A. and Beard, J.S. (1985). The fire factor in vine thicket and woodland vegetation of the Admiralty Gulf region, north-west Kimberley, Western Australia. *Proceedings of the Ecological Society of Australia* 13, 225–230.

Clements, A. (1983). Suburban development and resultant changes in the vegetation of the bushland of the northern Sydney region. *Australian Journal of Ecology* 8, 307–319.

Coaldrake, J.E. and Haydock, K.P. (1958). Soil phosphate and vegetal pattern in some natural communities of south-eastern Queensland, Australia. *Ecology* 39, 1–5.

Colhoun, E.A. and van de Geer, G. (1988). Darwin Crater, the King and Linda Valleys. In *Cainozoic Vegetation of Tasmania: Handbook prepared for 7th International Palynological Congress* (ed. E.A. Colhoun) pp. 30–71. (University of Newcastle: Newcastle, NSW.)

Collins, N.M., Sayer, J.A. and Whitmore, T.C. (1991). *The Conservation Atlas of Tropical Forests: Asia and the Pacific.* (Macmillan: London.)

Connell, J.H. and Slatyer, R.O. (1977). Mechanisms of succession in natural communities and their role in community stability and organization. *American Naturalist* 111, 1119–1144.

Connor, D.J. and Tunstall, B.R. (1968). Tissue water relations for brigalow and mulga. *Australian Journal of Botany* 16, 487–490.

Cosgrove, R. (1989). Thirty thousand years of human colonization in Tasmania: new Pleistocene dates. *Science* 243, 1706–1708.

Cosgrove, R. (1996). Origin and development of Australian Aboriginal tropical rainforest culture: a reconsideration. *Antiquity* 70, 900–912.

Cosgrove, R., Allen J. and Marshall, B. (1990). Palaeo-ecology and Pleistocene human occupation in south central Tasmania. *Antiquity* 64, 59–78.

Cosgrove, R., Allen, J. and Marshall, B. (1994). Late Pleistocene human occupation in Tasmania: a reply to Thomas. *Australian Archaeology* 38, 28–35.

Coventry, R.J., Stephenson, P.J. and Webb, A.W. (1985). Chronology of landscape evolution and soil development in the upper Flinders River area, Queensland, based on isotopic dating of Cainozoic basalts. *Australian Journal of Earth Sciences* 32, 433–447.

Cremer, K.W. (1960). Eucalypts in rain forest. *Australian Forestry* 24, 120–126.

Cremer, K.W. (1962). The effects of burnt soil on the growth rate of eucalypt seedlings. *Institute of Foresters of Australia Newsletter* 3, 2–5.

Cremer, K.W. (1965). Effects of fire on seedshed from *Eucalyptus regnans*. *Australian Forestry* 29, 252–262.

Cremer, K.W. and Mount, A.B. (1965). Early stages of plant succession following the complete felling and burning of *Eucalyptus regnans* forest in the Florentine Valley, Tasmania. *Australian Journal of Botany* 13, 303–322.

Crome, F.H.J. and Brown, H.E. (1979). Notes on social organisation and breeding of the Orange-footed Scrubfowl *Megapodius reinwardt. Emu* 79, 111–119.

Cromer, D.A.N., and Pryor, L.D. (1943). A contribution to rain-forest ecology. *Proceedings of the Linnean Society of New South Wales* 67, 249–268.

Cullen, P.J. (1987). Regeneration patterns in populations of *Athrotaxis selaginoides* D. Don. from Tasmania. *Journal of Biogeography* 14, 39–51.

Dale, M., Kershaw, P., Kikkawa, J., Parsons, P. and Webb, L. (1980). Resolution by the Ecological Society of Australia on Australian rainforest conservation. *Bulletin of the Ecological Society of Australia* 10, 6–7.

Davidson, N.J. and Reid, J.B. (1985). Frost as a factor influencing the growth and distribution of subalpine eucalypts. *Australian Journal of Botany* 33, 657–667.

Davis, C. (1936). Plant ecology of the Bulli district. *Proceedings of the Linnean Society of New South Wales* 61, 285–297.

Dickinson, K.J.M. and Kirkpatrick, J.B. (1985). The flammability and energy content of some important plant species and fuel components in the forests of southeastern Tasmania. *Journal of Biogeography* 12, 121–134.

Diels, L. (1906). *Die Vegetation der Erde VII. 'Die Pflanzenwelt von West Australien'.* (Engelmann: Leipzig.)

Dixon, K.W., Roche, S. and Pate, J.S. (1995). The promotive effect of smoke derived from burnt native vegetation on seed germination of Western Australian plants. *Oecologia* 101, 185–192.

Dodson, J.R. and Myers, C.A. (1986). Vegetation and modern pollen rain from the Barrington Tops and upper Hunter River regions of New South Wales. *Australian Journal of Botany* 34, 293–304.

Dodson, J., Greenwood, P.W. and Jones, R.L. (1986). Holocene forest and wetland vegetation dynamics at Barrington Tops, New South Wales. *Journal of Biogeography* 13, 561–585.

Dodson, J.R., McRae, V.M., Molloy, K., Roberts, F. and Smith, J.D. (1993). Late Holocene human impact on two coastal environments in New South Wales, Australia: a comparison of Aboriginal and European impacts. *Vegetation History and Archaeobotany* 2, 89–100.

Dodson, J.R., Roberts, F.K. and De Salis, T. (1994). Palaeoenvironments and human impact at Burraga Swamp in montane rainforest, Barrington Tops National Park, New South Wales, Australia. *Australian Geographer* 25 (2), 161–169.

Doley, D. (1978). Effects of shade on gas exchange and growth in seedlings of *Eucalyptus grandis* Hill ex Maiden. *Australian Journal of Plant Physiology* 5, 723–738.

Doley, D., Unwin, G.L. and Yates, D.J. (1988). Spatial and temporal distribution of photosynthesis and transpiration by single leaves in a rainforest tree, *Argyrodendron peralatum*. *Australian Journal of Plant Physiology* 15, 317–326.

Doley, D., Yates, D.J. and Unwin, G.L. (1987). Photosynthesis in an Australian rainforest tree, *Argyrodendron peralatum*, during the rapid development and relief of water deficits in the dry season. *Oecologia* 74, 441–449.

Domin, K. (1911). Queensland plant associations (some problems of Queensland's botanogeography). *Proceedings of the Royal Society of Queensland* 23, 57–74.

Dowe, J. (1995). A preliminary review of the biogeography of Australian Arecaceae. *Mooreana* 5, 7–22.

Dransfield, J. (1981). Palms and Wallace's line. In *Wallace's line and plate tectonics* (ed. T.C. Whitmore) pp. 43–56. (Clarendon Press: Oxford.)

Duff, G.A. and Stocker, G.C. (1989). The effects of frosts on rainforest/open forest ecotones in the highlands of north Queensland. *Proceedings of the Royal Society of Queensland* 100, 49–54.

Duggin, J.A. and Gentle, C.B. (1998). Experimental evidence on the importance of disturbance intensity for invasion of *Lantana camara* L. in dry rainforest–open forest ecotones in north-eastern NSW, Australia. *Forest Ecology and Management* 109, 279–292.

Dunlop, C.R. and Webb, L.J. (1991). Flora and vegetation. In *Monsoonal Australia: Landscape, ecology and man in the northern lowlands* (eds. C.D. Haynes, M.G. Ridpath and M.A.J. Williams) pp. 41–60. (A.A. Balkema: Rotterdam.)

Dunphy, M. (1991). Rainforest weeds of the Big Scrub. In *Rainforest remnants* (ed. S. Phillips) pp. 85–93. (NSW National Parks and Wildlife Service: Sydney.)

Dyne, G.R. (1991). Earthworm fauna of Australian rainforests. In *The rainforest legacy: Australian National Rainforests Study. Volume 2—Flora and fauna of the rainforests* (eds. G. Werren and P. Kershaw) pp. 335–343. (Australian Government Publishing Service: Canberra.)

Ellis, R.C. (1964). Dieback of alpine ash in north eastern Tasmania. *Australian Forestry* 27, 75–90.

Ellis, R.C. (1965). Correspondence. *Australian Forestry* 29, 56–57.

Ellis, R.C (1971a). Rainfall, fog drip and evaporation in a mountainous area of southern Australia. *Australian Forestry* 35, 99–106.

Ellis, R.C. (1971b). Dieback of alpine ash as related to changes in soil temperature. *Australian Forestry* 35, 152–163.

Ellis, R,C. (1985). The relationship among eucalypt forest, grassland and rainforest in a highland area in north-eastern Tasmania. *Australian Journal of Ecology* 10, 297–314.

Ellis, R.C. and Graley, A.M. (1983). Gains and losses in soil nutrients associated with harvesting and burning eucalypt rainforest. *Plant and Soil* 74, 437–450.

Ellis, R.C. and Graley, A.M. (1987). Soil chemical properties as related to forest succession in a highland area in north-east Tasmania. *Australian Journal of Ecology* 12, 307–317.

Ellis, R.C., Lowry, R.K. and Davies, S.K. (1982). The effect of regeneration burning upon the nutrient status of soil in two forest types in southern Tasmania. *Plant and Soil* 65, 171–186.

Ellis, R.C, Mount, A.B. and Mattay, J.P. (1980). Recovery of *Eucalyptus delegatensis* from high altitude dieback after felling and burning the understorey. *Australian Forestry* 43, 29–35.

Ellis, R.C. and Pennington, P.I. (1992). Factors affecting the growth of *Eucalyptus delegatensis* seedlings in inhibitory forest and grassland soils. *Plant and Soil* 145, 93–105.

Ellis, R.C and Thomas, I. (1988). Pre-settlement and post-settlement vegetational change and probable Aboriginal influences in a highland forested area in Tasmania. In *Australia's ever changing forests* (eds. K.J. Frawley and N.M. Semple) pp. 199–214. (Department of Geography and Oceanography, Australian Defence Force Academy: Canberra.)

Farrell, T.P. and Ashton, D.H. (1978). Populations studies on *Acacia melanoxylon* R.Br. I. Variation in seed and vegetative characteristics. *Australian Journal of Botany* 26, 365–379.

Feller, M.C. (1980). Biomass and nutrient distribution in two eucalypt forest ecosystems. *Australian Journal of Ecology* 5, 309–333.

Fensham, R.J. (1990). Interactive effects of fire frequency and site factors in tropical *Eucalyptus* forest. *Australian Journal of Ecology* 15, 255–266.

Fensham, R.J. (1993). The environmental relations of vegetation pattern on chenier ridges on Bathurst Island, Northern Territory. *Australian Journal of Botany* 41, 275–291.

Fensham, R.J. (1995). Floristics and environmental relations of inland dry rainforest in north Queensland, Australia. *Journal of Biogeography* 22, 1047–1063.

Fensham, R.J. (1996). Land clearance and conservation of inland dry rainforest in north Queensland, Australia. *Biological Conservation* 75, 289–298.

Fensham, R.J. (1997). Aboriginal fire regimes in Queensland, Australia: analysis of the explorers' record. *Journal of Biogeography* 24, 11–22.

Fensham, R.J. and Bowman, D.M.J.S. (1995). A comparison of foliar nutrient concentration in trees from monsoon rainforest and savanna in northern Australia. *Australian Journal of Ecology* 20, 335–339.

Fensham, R.J. and Fairfax, R.J. (1996). The disappearing grassy balds of the Bunya Mountains, south-eastern Queensland. *Australian Journal of Botany* 44, 543–558.

Fensham, R.J., Fairfax, R.J. and Cannell, R.J. (1994). The invasion of *Lantana camara* L. in Forty Mile Scrub National Park, north Queensland. *Australian Journal of Ecology* 19, 297–305.

Fensham, R.J. and Kirkpatrick, J.B. (1992). Soil characteristics and tree species distributions in the savannah of Melville Island, Northern Territory. *Australian Journal of Botany* 40, 311–333.

Fisher, H.J. (1985). The structure and floristic composition of the rainforest of the Liverpool Range, New South Wales, and its relationships with other Australian rainforests. *Australian Journal of Ecology* 10, 315–325.

Flannery, T.F. (1989). Phylogeny of the Macropodoidea; a study in convergence. In *Kangaroos, Wallabies and Rat-kangaroos* (eds. G. Grigg, P. Jarman and I. Hume) pp. 1–46. (Surrey Beatty and Sons: Sydney.)

Flannery, T. (1990*a*). Who killed kirlilpi? *Australian Natural History* 23, 234–241.

Flannery, T.F. (1990*b*). Pleistocene faunal loss: implications of the aftershock for Australia's past and future. *Archaeology in Oceania* 25, 45–67.

Flannery, T.F., Boeadi and Szalay, A.L. (1995). A new tree-kangaroo (*Dendrolagus*: Marsupialia) from Irian Jaya, Indonesia, with notes on ethnography and the evolution of tree-kangaroos. *Mammalia* 59, 65–84.

Flohn, H. (1973). Antarctica and the global Cenozoic evolution: a geophysical model. In *Palaeoecology of Africa & of the surrounding islands & Antarctica* (ed. E.M. van Zinderen Bakker) pp. 37–53. (A.A. Balkema: Cape Town.)

Florence, R.G. (1964). Edaphic control of vegetational pattern in east coast forests. *Proceedings of the Linnean Society of New South Wales* 84, 171–190.

Florence, R.G. (1965). Some vegetation-soil relationships in the Blackall Range forests. *Australian Forestry* 29, 105–118.

Florence, R.G. and Crocker, R.L. (1962). Analysis of blackbutt (*Eucalyptus pilularis* SM.) seedling growth in a blackbutt forest soil. *Ecology* 43, 670–679.

Floyd, A.G. (1990). *Australian Rainforests in New South Wales. Volume 1.* (Surrey Beatty and Sons: New South Wales.)

Floyd, A.G. (1991). Natural regeneration of rainforest in N.S.W. In *Rainforest Remnants* (ed. S. Phillips) pp. 20–25. (NSW National Parks and Wildlife Service: Sydney.)

Fordyce, I.R., Duff, G.A. and Eamus, D. (1995). The ecophysiology of *Allosyncarpia ternata* (Myrtaceae) in northern Australia: tree physiognomy, leaf characteristics and assimilation at contrasting sites. *Australian Journal of Botany* 43, 367–377.

Fordyce, I.R., Duff, G.A. and Eamus, D. (1997*a*). The water relations of *Allosyncarpia ternata* (Myrtaceae) at contrasting sites in the monsoonal tropics of northern Australia. *Australian Journal of Botany* 45, 259–274.

Fordyce, I.R., Eamus, D., Duff, G.A. and Williams, R.J. (1997*b*). The role of seedling age and size in the recovery of *Allosyncarpia ternata* following fire. *Australian Journal of Ecology* 22, 262–269.

Francis, W.D. (1951). *Australian rain-forest trees.* (Forestry and Timber Bureau: Canberra.)

Fraser, L. and Vickery, J.W. (1937). The ecology of the upper Williams River and Barrington Tops Districts. I. Introduction. *Proceedings of the Linnean Society of New South Wales* 62, 269–283.

Fraser, L. and Vickery, J.W. (1938). The ecology of the upper Williams River and Barrington Tops Districts. II. The rain-forest formations. *Proceedings of the Linnean Society of New South Wales* 63, 139–184.

Fraser, L. and Vickery, J.W. (1939). The ecology of the upper Williams River and Barrington Tops. III. The eucalypt forests, and general discussion. *Proceedings of the Linnean Society of New South Wales* 64, 1–33.

Freeland, W.J. and Winter, J.W. (1975). Evolutionary consequences of eating: *Trichosurus vulpecula* (Marsupialia) and the genus *Eucalyptus*. *Journal of Chemical Ecology* 1, 439–455.

Fullagar, R.L.K., Price, D.M. and Head, L.M. (1996). Early human occupation of northern Australia: archaeology and thermoluminescence dating of Jinmium rock-shelter, Northern Territory. *Antiquity* 70, 751–773.

Furley, P.A., Proctor, J. and Ratter, J.A. (1992). *Nature and dynamics of forest– savanna boundaries*. (Chapman and Hall: London.)

Gell, P. and Mercer, D. (1992). Introduction. In *Victoria's rainforests: perspectives on definition, classification and management* (eds. P. Gell and D. Mercer) pp. 1–8. (Monash Publications in Geography No. 41: Melbourne.)

Gell, P.A., Stuart, I. and Smith, J.D. (1993). The response of vegetation to changing fire regimes and human activity in east Gippsland, Victoria, Australia. *Holocene* 3, 150–160.

Gentle, C.B. and Duggin, J.A. (1997). *Lantana camara* L. invasions in dry rainforest—open forest ecotones: the role of disturbances associated with fire and cattle grazing. *Australian Journal of Ecology* 22, 298–306.

Gibson, N. and Brown, M.J. (1991). The ecology of *Lagarostrobos franklinii* (Hook.f.) Quinn (Podocarpaceae) in Tasmania. 2. Population structure and spatial pattern. *Australian Journal of Ecology* 16, 223–229.

Gibson, N., Davies, J. and Brown, M.J. (1991). The ecology of *Lagarostrobos franklinii* (Hook.f.) Quinn (Podocarpaceae) in Tasmania. 1. Distribution, floristics and environmental correlates. *Australian Journal of Ecology* 16, 215–222.

Gilbert, J.M. (1959). Forest succession in the Florentine Valley, Tasmania. *Papers and Proceedings of the Royal Society of Tasmania* 93, 129–151.

Gill, A.M., Groves, R.H. and Noble, I.R. (1981). *Fire and the Australian biota*. (Australian Academy of Sciences: Canberra.)

Gill, A.M., Moore, P.H.R. and Williams, R.J. (1996). Fire weather in the wet-dry tropics of the World Heritage Kakadu National Park, Australia. *Australian Journal of Ecology* 21, 302–308.

Gillison, A.N. (1981). Towards a functional vegetation classification. In *Vegetation Classification in Australia* (eds. A.N. Gillison and D.J. Anderson) pp. 30–41. (Commonwealth Scientific and Industrial Research Organisation and Australian National University: Canberra.)

Gillison, A.N. (1983). Tropical savannas of Australia and the southwest Pacific. In *Ecosystems of the World. Volume 13: Tropical Savannas* (ed. F. Bourliere) pp. 183–243. (Elsevier Scientific Publishing: Amsterdam.)

Gleadow, R.M. (1982). Invasion by *Pittosporum undulatum* of the forests of central Victoria. II Dispersal, germination and establishment. *Australian Journal of Botany* 30, 185–198.

Gleadow, R.M. and Ashton, D.H. (1981). Invasion by *Pittosporum undulatum* of the forests of central Victoria. I Invasion patterns and plant morphology. *Australian Journal of Botany* 29, 705–720.

Gleadow, R.M. and Rowan, K.S. (1982). Invasion by *Pittosporum undulatum* of the forests of central Victoria. III Effects of temperature and light on growth and drought resistance. *Australian Journal of Botany* 30, 347–357.

Gleadow, R.M., Rowan, K.S. and Ashton, D.H. (1983). Invasion by *Pittosporum undulatum* of the forests of central Victoria. IV Shade tolerance. *Australian Journal of Botany* 31, 151–160.

Golson, J. (1971). Australian Aboriginal food plants: some ecological and culture-historical implications. In *Aboriginal man and environment in Australia* (eds. D.J. Mulvaney and J. Golson) pp. 196–238. (Australian National University Press: Canberra.)

Gould, R.A. (1971). Uses and effects of fire among the western desert Aborigines of Australia. *Mankind* 8, 14–24.

Green, D.G. (1989). Simulated effects of fire, dispersal and spatial pattern on competition within forest mosaics. *Vegetatio* 82, 139–153.

Greenwood, D.R. (1994). Palaeobotanical evidence for Tertiary climates. In *History of the Australian vegetation: Cretaceous to Recent* (ed. R.S. Hill) pp. 44–59. (Cambridge University Press: Cambridge.)

Grieve, B.J. (1955). The physiology of sclerophyll plants. *Journal of the Royal Society of Western Australia* 39, 31–45.

Grose, M.J. (1989). Phosphorus nutrition of seedlings of the Waratah, *Telopea speciosissima* (Sm.) R.Br. (Proteaceae). *Australian Journal of Botany* 37, 313–320.

Groves, R.H. and Keraitis, K. (1976). Survival and growth of seedlings of three sclerophyll species at high levels of phosphorus and nitrogen. *Australian Journal of Botany* 24, 681–690.

Grundon, N.J. (1972). Mineral nutrition of some Queensland heath plants. *Journal of Ecology* 60, 171–181.

Hallam, S.J. (1975). *Fire and hearth: a study of Aboriginal usage and European usurpation in south-western Australia.* (Australian Institute of Aboriginal Studies: Canberra.)

Hallam, S.J. (1985). The history of Aboriginal firing. In *Fire Ecology and management in Western Australian ecosystems* (ed. J.R. Ford) pp. 7–20. (Western Australian Institute of Technology: Perth.)

Hamilton, A.G. (1914). The xerophilous characters of *Hakea dactyloides* Cav. [N.O. Proteaceae]. *Proceedings of the Linnean Society of New South Wales* 39, 152–156.

Harms, J.E., Milton, D.J., Ferguson, J., Gilbert, D.J., Harris, W.K. and Goleby, B. (1980). Goat paddock cryptoexplosion crater, Western Australia. *Nature* 286, 704–706.

Harrington, G. (1995). Should we play God with the rainforest? *Wildlife Australia,* Winter 1995, 8–11.

Harrington, G.N. and Sanderson, K.D. (1994). Recent contraction of wet sclerophyll forest in the wet tropics of Queensland due to invasion by rainforest.
Pacific Conservation Biology 1, 319–327.

Harwood, C.E. (1980). Frost resistance of subalpine *Eucalyptus* species. I. Experiments using a radiation frost room. *Australian Journal of Botany* 28, 587–599.

Harwood, C.E. and Jackson, W.D. (1975). Atmospheric losses of four plant
nutrients during a forest fire. *Australian Forestry* 38, 92–99.

Haynes, C.D. (1978). Land, trees and man (*Gunret, gundulk, dja bining*).
Commonwealth Forestry Review 57, 99–106.

Head, L. (1989). Prehistoric Aboriginal impacts on Australian vegetation: an
assessment of the evidence. *Australian Geographer* 20, 37–46.

Henderson, W. and Wilkins, C.W. (1975). The interaction of bushfires and vegetation. *Search* 6, 130–133.

Herbert, D.A. (1929). The major factors in the present distribution of the genus
Eucalyptus. *Proceedings of the Royal Society of Queensland* 40, 165–193.

Herbert, D.A. (1932). The relationships of the Queensland flora. *Proceedings of the
Royal Society of Queensland* 44, 2–22.

Herbert, D.A. (1936). An advancing Antarctic Beech forest. *Queensland Naturalist*
10, 8–10.

Herbert, D.A. (1938). The upland savannahs of the Bunya Mountains, South
Queensland. *Proceedings of the Royal Society of Queensland* 49, 145–149.

Herbert, D.A. (1960). Tropical and sub-tropical rain forest in Australia. *Australian
Journal of Science* 22, 283–290.

Hiatt, B. (1968). The food quest and the economy of the Tasmanian Aborigines.
Oceania 38, 99–133, 190–219.

Hill, R.S. (1982). Rainforest fire in western Tasmania. *Australian Journal of Botany*
30, 583–589.

Hill, R.S. (1983). Evolution of *Nothofagus cunninghamii* and its relationship to *N.
moorei* as inferred from Tasmania macrofossils. *Australian Journal of Botany*
31, 453–465.

Hill, R.S. (1994a). *History of the Australian vegetation: Cretaceous to Recent.* (Cambridge University Press: Cambridge.)

Hill, R.S. (1994b). The Australian fossil plant record: an introduction. In *History of
the Australian vegetation: Cretaceous to Recent* (ed. R.S. Hill) pp. 1–4. (Cambridge University Press: Cambridge.)

Hill, R.S. (1994c). The history of selected Australian taxa. In *History of the Australian vegetation: Cretaceous to Recent* (ed. R.S. Hill) pp. 390–419. (Cambridge
University Press: Cambridge.)

Hill, R.S. and Read, J. (1984). Post-fire regeneration of rainforest and mixed forest
in western Tasmania. *Australian Journal of Botany* 32, 481–493.

Hill, R.S. and Read, J. (1987). Endemism in Tasmanian cool temperate rainforest:
alternative hypotheses. *Botanical Journal of the Linnean Society* 95, 113–124.

Hill, R.S., Read, J. and Busby, J.R. (1988). The temperature-dependence of photosynthesis of some Australian temperate rainforest trees and its biogeographi-

cal significance. *Journal of Biogeography* 15, 431–449.

Hiscock, P. (1990). How old are the artefacts at Malakunanja II. *Archaeology in Oceania* 25, 122–124.

Hnatiuk, R.J. (1977). Population structure of *Livistona eastonii* Gardn., Mitchell Plateau, Western Australia. *Australian Journal of Ecology* 2, 461–466.

Hnatiuk, R.J. and Kenneally, K.F. (1981). A survey of the vegetation and flora of Mitchell Plateau, Kimberley, Western Australia. In *Biological Survey of Mitchell Plateau and Admiralty Gulf, Kimberley, Western Australia* pp. 13–94. (Western Australian Museum: Perth.)

Hocking, P.J. (1982). The nutrition of fruits of two proteaceous shrubs, *Grevillea wilsonii* and *Hakea undulata*, from south-western Australia. *Australian Journal of Botany* 30, 219–230.

Hocking, P.J. (1986). Mineral nutrient composition of leaves and fruits of selected species of *Grevillea* from south-western Australia, with special reference to *Grevillea leucopteris* Meissn. *Australian Journal of Botany* 34, 155–164.

Holdridge, L.R. (1947). Determination of world plant formations from simple climatic data. *Science* 105, 367–368.

Hooker, J.D. (1860). *The Botany of the Antarctic Voyage of H.M. Discovery Ships* Erebus *and* Terror *in the years 1839–1843. Part III. Flora Tasmaniae. Vol. I. Dicotyledones.* (Reeve: London.)

Hope, G.S. (1994a). Comment on ODP site 820 and the inference of early human occupation in Australia. *Quaternary Australasia* 12, 32–33.

Hope, G.S. (1994b). Quaternary vegetation. In *History of the Australian vegetation: Cretaceous to Recent* (ed. R.S. Hill) pp. 368–389. (Cambridge University Press: Cambridge.)

Hope, G.S. (1999). Vegetation and fire response to late Holocene human occupation in island and mainland north west Tasmania. *Quaternary International* 59, 47–60.

Hopkins, M.S., Ash, J., Graham, A.W., Head, J. and Hewett, R.K. (1993). Charcoal evidence of the spatial extent of the *Eucalyptus* woodland expansions and rainforest contractions in North Queensland during the late Pleistocene. *Journal of Biogeography* 20, 357–372.

Hopkins, M.S. and Graham, A.W. (1984a). Viable soil seed banks in disturbed lowland tropical rainforest sites in North Queensland. *Australian Journal of Ecology* 9, 71–79.

Hopkins, M.S. and Graham, A.W. (1984b). The role of soil seed banks in regeneration in canopy gaps in Australian tropical lowland rainforest – preliminary field experiments. *Malaysian Forester* 47, 146–158.

Hopkins, M.S. and Graham, A.W. (1987). Gregarious flowering in a lowland tropical rainforest: A possible response to disturbance by Cyclone Winifred. *Australian Journal of Ecology* 12, 25–29.

Hopkins, M.S., Head, J., Ash, J., Hewett, R.K. and Graham A.W. (1996). Evidence of a Holocene and continuing recent expansion of lowland rain forest in humid, tropical North Queensland. *Journal of Biogeography* 23: 737–745.

Hopkins, M.S., Tracey, J.G. and Graham, A.W. (1990). The size and composition of soil seed-banks in remnant patches of three structural rainforest types in North Queensland. *Australian Journal of Ecology* 15, 43–50.

Hopper, S.D. (1979). Biogeographical aspects of speciation in the southwest Australian flora. *Annual Review of Ecology and Systematics* 10, 399–422.

Horton, D.R. (1982). The burning question: Aborigines, fire and Australian ecosystems. *Mankind* 13, 237–251.

Horton, D.R. (1984). Red kangaroos: last of the Australian megafauna. In *Quaternary extinctions: a prehistoric revolution* (eds. P.S. Martin and R.G. Klein) pp. 639–680. (The University of Arizona Press: Tucson.)

Howard, T.M. (1973*a*). Studies in the ecology of *Nothofagus cunninghamii* Oerst. I. Natural regeneration on the Mt. Donna Buang massif, Victoria. *Australian Journal of Botany* 21, 67–78.

Howard, T.M. (1973*b*). Studies in the ecology of *Nothofagus cunninghamii* Oerst. III. Two limiting factors: light intensity and water stress. *Australian Journal of Botany* 21, 93–102.

Howard, T.M. and Ashton, D.H. (1973). The distribution of *Nothofagus cunninghamii* rainforest. *Proceedings of the Royal Society of Victoria* 85, 47–75.

Hughes, L., Cawsey, E.M. and Westoby, M. (1996). Climatic range sizes of *Eucalyptus* species in relation to future climate change. *Global Ecology and Biogeography Letters* 5, 23–29.

Humphreys, W.F., Brooks, R.D. and Vine, B. (1990). Rediscovery of the palm *Livistona alfredii* on the North West Cape Peninsula. *Records of the West Australian Museum* 14, 647–650.

Hutley, L.B., Doley, D., Yates, D.J. and Boonsaner, A. (1997). Water balance of an Australian subtropical rainforest at altitude: the ecological and physiological significance of intercepted cloud and fog. *Australian Journal of Botany* 45, 311–329.

Jackson, W.D. (1965). Vegetation. In *Atlas of Tasmania* (ed. J.L. Davies) pp. 30–35. (Lands and Survey Department: Hobart.)

Jackson, W.D. (1968). Fire, air, water and earth—An elemental ecology of Tasmania. *Proceedings of the Ecological Society of Australia* 3, 9–16.

Jackson, W.D. (1999). The Tasmanian legacy of man and fire. *Papers and Proceedings of the Royal Society of Tasmania* 133, 1–14.

Jackson, W.D. and Bowman, D.M.J.S. (1982*a*). Reply: Ecological drift or fire cycles in south-west Tasmania. *Search* 13, 175–176.

Jackson, W.D. and Bowman, D.M.J.S. (1982*b*). Comment: slashburning in Tasmanian dry eucalypt forests. *Australian Forestry* 45, 63–67.

Janzen, D.H. (1974). Tropical blackwater rivers, animals, and mast fruiting by the Dipterocarpaceae. *Biotropica* 6, 69–103.

Janzen D.H. (1988). Complexity is in the eye of the beholder. In *Tropical Rainforests: diversity and conservation* (eds. F. Almeda and C.M. Pringle) pp. 29–51. (California Academy of Sciences and Pacific Division, American Association for the Advancement of Science: San Francisco.)

Jarman, S.J. and Brown, M.J. (1983). A definition of cool temperate rainforest in Tasmania. *Search* 14, 81–87.

Jarman, S.J., Brown, M.J. and Kantvilas, G. (1987). The classification, distribution and conservation status of Tasmanian rainforests. In *The rainforest legacy: Australian national rainforests study. Volume 1—the nature, distribution and status of rainforest types* (eds G. Werren and A.P. Kershaw) pp. 9–22. (Australian Government Publishing Service: Canberra.)

Jarrett, P.H. and Petrie, A.H.K. (1929). The vegetation of the Blacks' Spur region: a study in the ecology of some Australian mountain *Eucalyptus* forests II. Pyric succession. *Journal of Ecology* 17, 249–281.

Jennings, J.N. and Mabbutt, J.A. (1986). Introduction. In *Australia, a geography. Vol. 1. The natural environment* (ed. D.J. Jeans) pp. 80–96. (Sydney University Press: Sydney.)

Johnson, C.N. (1996). Interactions between mammals and ectomycorrhizal fungi. *Trends in Ecology and Evolution* 11, 503–507.

Johnson, K.A. (1980). Spatial and temporal use of habitat by the red-necked pademelon, *Thylogale thetis* (Marsupialia: Macropodidae). *Australian Wildlife Research* 7, 157–166.

Johnson, L.A.S. and Briggs, B.G. (1981). Three old southern families—Myrtaceae, Proteaceae and Restionaceae. In *Ecological biogeography of Australia* (ed. A. Keast) pp. 427–469. (Dr. W. Junk: The Hague.)

Johnston, R.D. and Lacey, C.J. (1983). Multi-stemmed trees in rainforest. *Australia Journal of Botany* 31, 189–195.

Johnston, R.D. and Lacey, C.J. (1984). A proposal for the classification of tree-dominated vegetation in Australia. *Australian Journal of Botany* 32, 529–549.

Jones, R. (1969). Fire-stick farming. *Australian Natural History* 16, 224–228.

Jones, R. (1975). The Neolithic, Palaeolithic and the hunting gardeners: Man and land in the Antipodes. In *Quaternary studies* (eds. R.P. Suggate and M.M. Cresswell) pp. 21–34. (Royal Society of New Zealand: Wellington.)

Jones, R. (1979). The fifth continent: problems concerning the human colonization of Australia. *Annual Review of Anthropology* 8, 445–466.

Jones, R. (1980). Hunters in the Australian coastal savanna. In *Human ecology in savanna environments* (ed. D.R. Harris) pp. 107–146. (Academic Press: London.)

Judd, T.S. (1994). Do small myrtaceous seed-capsules display specialized insulating characteristics which protect seed during fire? *Annals of Botany* 73, 33–38.

Keast, A. (1959). Relict animals and plants of the Macdonnell Ranges. *Australian Museum Magazine* 13, 81–86.

Kemp, E.M. (1978). Tertiary climatic evolution and vegetation history in the southeast Indian Ocean region. *Palaeogeography, Palaeoclimatology, Palaeoecology* 24, 169–208.

Kemp, E.M. (1981). Pre-Quaternary fire in Australia. In *Fire and the Australian biota* (eds. A.M. Gill, R.H. Groves and I.R. Noble) pp. 3–21. (Australian Academy of Science: Canberra.)

Kershaw, A.P. (1974). A long continuous pollen sequence from north-eastern Australia. *Nature* **251**, 222–223.

Kershaw, A.P. (1976). A late Pleistocene and Holocene pollen diagram from Lynch's Crater, north-eastern Queensland, Australia. *New Phytologist* **77**, 469–498.

Kershaw, A.P. (1978). Record of last interglacial-glacial cycle from northeastern Queensland. *Nature* **272**, 159–161.

Kershaw, A.P. (1981). Quaternary vegetation and environments. In *Ecological biogeography of Australia* (ed. A. Keast) pp. 83–101. (Dr W. Junk: The Hague.)

Kershaw, A.P. (1984). Late Cenozoic plant extinctions in Australia. In *Quaternary extinctions: a prehistoric revolution* (eds. P.S. Martin and R.G. Klein) pp. 691–707. (University of Arizona Press: Tucson.)

Kershaw, A.P. (1985). An extended late Quaternary vegetation record from north-eastern Queensland and its implications for the seasonal tropics of Australia. *Proceedings of the Ecological Society of Australia* **13**, 179–189.

Kershaw, A.P. (1986). Climatic change and Aboriginal burning in north-east Australia during the last two glacial/interglacial cycles. *Nature* **322**, 47–49.

Kershaw, A.P. (1994*a*). Site 820 and the evidence for early occupation in Australia—a response. *Quaternary Australasia* **12**, 24–29.

Kershaw A.P. (1994*b*). Pleistocene vegetation of the humid tropics of northeastern Queensland, Australia. *Palaeogeography, Palaeoclimatology, Palaeoecology* **109**, 399–412.

Kershaw, A.P., D'Costa, D.M., McEwen, J., Mason, J.R.C. and Wagstaff, B.E. (1991*a*). Palynological evidence for Quaternary vegetation and environments of mainland southeastern Australia. *Quaternary Science Reviews* **10**, 391–404.

Kershaw, A.P., McKenzie, G.M. and McMinn, A. (1993). A Quaternary vegetation history of northeastern Queensland from pollen analysis of ODP Site 820. *Proceedings of the Ocean Drilling Program, Scientific Results* **133**, 107–114.

Kershaw, A.P., Martin, H.A. and McEwen Mason, J.R.C. (1994). The Neogene: a period of transition. In *History of the Australian vegetation: Cretaceous to Recent* (ed. R.S. Hill) pp. 299–327. (Cambridge University Press: Cambridge.)

Kershaw, A.P. and Sluiter, I.R. (1982). Late Cenozoic pollen spectra from the Atherton Tableland, north-eastern Australia. *Australian Journal of Botany* **30**, 279–295.

Kershaw, A.P., Sluiter, I.R., McEwan Mason, J., Wagstaff, B.E. and Whitelaw, M. (1991*b*). The history of rainforest in Australia—evidence from pollen. In *The rainforest legacy: Australian National Rainforests Study Volume 3—Rainforest history, dynamics and management* (eds. G. Werren and P. Kershaw) pp. 1–15. (Australian Government Publishing Service: Canberra.)

Kimber, R. (1983). Black lightning: Aborigines and fire in Central Australia and the Western Desert. *Archaeology in Oceania* **18**, 38–44.

King, A.R. (1963). *Report on the influence of colonization on the forests and the prevalence of bushfires in Australia*. (Commonwealth Scientific and Industrial Research Organisation Division of Physical Chemistry Mimeo Report: Melbourne.)

King, N.K. and Vines, R.G. (1969). *Variation in the flammability of the leaves of some Australian forest species.* (Commonwealth Scientific and Industrial Research Organisation Division of Applied Chemistry Memo: Melbourne.)

Kirkpatrick, J.B. (1977). Native vegetation of the west coast region of Tasmania. In *Landscape and Man. The interaction between man and the environment in western Tasmania* (eds. M.R. Banks and J.B. Kirkpatrick) pp. 55–79. (Royal Society of Tasmania: Hobart.)

Kirkpatrick, J. (1994). *A continent transformed: human impact on the natural vegetation of Australia.* (Oxford University Press: Melbourne.)

Kirkpatrick, J.B., Fensham, R.J., Nunez, M. and Bowman, D.M.J.S. (1988). Vegetation-radiation relationships in the wet-dry tropics: granite hills in northern Australia. *Vegetatio* **76**, 103–112.

Kirkpatrick, J., Meredith, C., Norton, T., Plumwood, V. and Fensham, R. (1990). *The Ecological Future of Australia's Forests.* (Australian Conservation Foundation: Melbourne.)

Kuuluvainen, T. (1992). Tree architecture adapted to efficient light utilization: is there a basis for latitudinal gradients? *Oikos* **65**, 275–284.

Lacey, C.J. and Jahnke, R, (1984). The occurrence and nature of lignotubers in *Notelaea longifolia* and *Elaeocarpus reticulatus. Australia Journal of Botany* **32**, 311–321.

Ladd, P.G. (1988). The status of Casuarinaceae in Australian forests. In *Australia's ever changing forests* (eds. K.J. Frawley and N.M. Semple) pp. 63–85. (Department of Geography and Oceanography, Australian Defence Force Academy, Special Publication No. 1: Campbell, ACT.)

Ladiges, P.Y. (1997). Phylogenetic history and classification of eucalypts. In *Eucalypt ecology: individuals to ecosystems* (eds. J. E. Williams and J.C.Z. Woinarski) pp.16–29. (Cambridge University Press: Cambridge.)

Ladiges, P.Y, Udovicic, F. and Drinnan, A.N. (1995). Eucalypt phylogeny – molecules and morphology. *Australian Systematic Botany* **8**, 483–497.

Lambert, M.J. and Turner, J. (1983). Soil nutrient-vegetation relationships in the Eden area, N.S.W. III. Foliage nutrient relationships with particular reference to *Eucalyptus* subgenera. *Australian Forestry* **46**, 200–209.

Lambert, M.J. and Turner, J. (1986). Nutrient concentrations in foliage of species within a New South Wales sub-tropical rainforest. *Annals of Botany* **58**, 465–478.

Lambert, M.J. and Turner, J. (1987). Suburban development and change in vegetation nutritional status. *Australian Journal of Ecology* **12**, 193–196.

Lambert, M.J. and Turner, J. (1989). Redistribution of nutrients in subtropical rainforest trees. *Proceedings of the Linnean Society of New South Wales* **111**, 1–10.

Lambert, M.J., Turner, J. and Kelly, J. (1983). Nutrient relationships of tree species in a New South Wales subtropical rainforest. *Australian Forest Research* **13**, 91–102.

Landsberg, J.J. and Cork, S.J. (1997). Herbivory: interaction between eucalypts and the vertebrates and invertebrates that feed on them. In *Eucalypt ecology: individuals to ecosystems* (eds. J.E. Williams and J.C.Z. Woinarski) pp. 342–372. (Cambridge University Press: Cambridge.)

Landsberg, J. and Ohmart, C. (1989). Levels of insect defoliation in forests: patterns and concepts. *Trends in Ecology and Evolution* 4, 96–100.

Langenheim, J.H., Osmond, C.B., Brooks, A. and Ferrar, P.J. (1984). Photosynthetic responses to light in seedlings of selected Amazonian and Australian rainforest tree species. *Oecologia* 63, 215–224.

Langkamp, P.J., Ashton, D.H. and Dalling, M.J. (1981). Ecological gradients in forest communities on Groote Eylandt, Northern Territory, Australia. *Vegetatio* 48, 27–46.

Langkamp, P.J. and Dalling, M.J. (1982). Nutrient cycling in a stand of *Acacia holosericea* A. Cunn. ex Don. II Phosphorus and endomycorrhizal associations. *Australian Journal of Botany* 30, 107–119.

Langkamp, P.J., Farnell, G.K. and Dalling, M.J. (1982). Nutrient cycling in a stand of *Acacia holosericea* A. Cunn. ex G. Don. I. Measurements of precipitation interception, seasonal acetylene reduction, plant growth and nitrogen requirement. *Australian Journal of Botany* 30, 87–106.

Latz, P.K. (1975). Notes on the relict palm *Livistona mariae* F. Muell. in central Australia. *Transactions of the Royal Society of South Australia* 99, 189–196.

Latz, P.K. and Griffin, G.F. (1978). Changes in Aboriginal land management in relation to fire and to food plants in central Australia. In *The nutrition of Aborigines in relation to the ecosystems of central Australia* (eds. B.S. Hetzel and H.J. Frith) pp. 77–85. (Commonwealth Scientific and Industrial Research Organisation: Melbourne.)

Le Brocque, A.F. and Buckney, R.T. (1994). Vegetation and environmental patterns on soils derived from Hawkesbury Sandstone and Narrabeen substrata in Ku-ring-gai Chase National Park, New South Wales. *Australian Journal of Ecology* 19, 229–238.

Lewis H.T. (1982). Fire technology and resource management in Aboriginal North America and Australia. In *Resource managers: North American and Australian hunter-gatherers* (eds. N.M. Williams and E.S. Hunn) pp. 45–67. (AAAS Selected Symposium Series No. 67, Westview Press: Colorado.)

Liddle, D.T., Russell-Smith, J., Brock, J., Leach, G.J. and Connors, G.T. (1994). *Atlas of the vascular rainforest plants of the Northern Territory.* (Flora of Australia Supplementary Series Number 3, Australian Biological Resources Study: Canberra.)

Loneragan, O.W. and Loneragan, J.F. (1964). Ashbed and nutrients in the growth of seedlings of Karri (*Eucalyptus diversicolor* F.v.M.). *Journal of the Royal Society of Western Australia* 47, 75–80.

Loveless, A.R. (1961). A nutritional interpretation of sclerophylly based on differences in the chemical composition of sclerophyllous and mesophytic leaves. *Annals of Botany* 25, 168–184.

Loveless, A.R. (1962). Further evidence to support a nutritional interpretation of sclerophylly. *Annals of Botany* 26, 551–561.

Lowe, I. (ed.). (1996). *Australia: State of the environment.* (Commonwealth Scientific and Industrial Research Organisation: Melbourne.)

Lowman, M.D. (1986). Light interception and its relation to structural differences in three Australian rainforest canopies. *Australian Journal of Ecology* 11, 163–170.

Lowman, M.D. and Box, J.D. (1983). Variation in leaf toughness and phenolic content among five species of Australian rain forest trees. *Australian Journal of Ecology* 8, 17–25.

Luke, R.H. and McArthur, A.G. (1978). *Bushfires in Australia.* (Australian Government Printing Service: Canberra.)

Macauley, B.J. and Fox, L.R. (1980). Variation in total phenols and condensed tannins in *Eucalyptus*: leaf phenology and insect grazing. *Australian Journal of Ecology* 5, 31–35.

Mackey, B.G., Nix, H.A., Stein, J.A., Cork, S.E. and Bullen, F.T. (1989). Assessing the representativeness of the wet tropics of Queensland World Heritage property. *Biological Conservation* 50, 279–303.

Macphail, M. (1975). Late Pleistocene environments in Tasmania. *Search* 6, 295–300.

Macphail, M.K. (1979). Holocene climatic change and Aboriginal food economy in Tasmania. *Search* 10, 11–12.

Macphail, M.K. (1980). Regeneration processes in Tasmanian forests: A long-term perspective based on pollen analysis. *Search* 11, 184–190.

Macphail, M.K. (1984). Small-scale dynamics in an early Holocene wet sclerophyll forest in Tasmania. *New Phytologist* 96, 131–147.

Macphail, M.K. (1991). Cool temperate rainforest: the not quite immemorial forest in Tasmania. In *The rainforest legacy: Australian National Rainforests Study. Volume 3—Rainforest history, dynamics and management* (eds. G. Werren and P. Kershaw) pp. 45–53. (Australian Government Publishing Service: Canberra.)

Macphail, M.K. and Colhoun, E.A. (1985). Late last glacial vegetation, climates and fire activity in southwest Tasmania. *Search* 16, 43–45.

Macphail, M.K., Jordon, G.J. and Hill, R.S. (1993). Key periods in the evolution of the flora and vegetation in western Tasmania I. the early-middle Pleistocene. *Australian Journal of Botany* 41, 673–707.

Malingreau, J.P., Stephens, G. and Fellows, L. (1985). Remote sensing of forest fires: Kalimantan and north Borneo in 1982–83. *Ambio* 14, 314–321.

Marsden-Smedley, J.B. and Catchpole, W.R. (1995a). Fire behaviour modelling in Tasmanian buttongrass moorlands I. Fuel characteristics. *International Journal of Wildland Fire* 5, 203–214.

Marsden-Smedley, J.B. and Catchpole, W.R. (1995b). Fire behaviour modelling in Tasmanian buttongrass moorlands II. Fire behaviour. *International Journal of Wildland Fire* 5, 215–228.

Martin, A.R.H. (1982). Proteaceae and the early differentiation of the central Australian flora. In *Evolution of the flora and fauna of arid Australia* (eds. W.R. Barker and P.J.M. Greenslade) pp. 77–83. (Peacock Publications: Adelaide.)

Martin, H.A. (1977). The history of *Ilex* (Aquifoliaceae) with special reference to Australia: evidence from pollen. *Australian Journal of Botany* 25, 655–673.

Martin, H.A. (1978). Evolution of the Australian flora and vegetation through the Tertiary: evidence from pollen. *Alcheringa* 2, 181–202.

Martin, H.A. (1987). Cainozoic history of the vegetation and climate of the Lachlan River region, New South Wales. *Proceedings of the Linnean Society of New South Wales* 109, 213–257.

Martin, H.A. (1994). Australian Tertiary phytogeography: evidence from palynology. In *History of the Australian vegetation: Cretaceous to Recent* (ed. R.S. Hill) pp. 104–142. (Cambridge University Press: Cambridge.)

Martin, H.A. and Specht, R.L. (1962). Are mesic communities less drought-resistant? A study on moisture relationships in dry sclerophyll forest at Inglewood, South Australia. *Australian Journal of Botany* 10, 106–118.

McColl, J.G. and Humpreys, F.R. (1967). Relationships between some nutritional factors and the distributions of *Eucalyptus gummifera* and *Eucalyptus maculata*. *Ecology* 48, 766–771.

McLuckie, J. and Petrie, A.H.K (1927). An ecological study of the flora of Mount Wilson. *Proceedings of the Linnean Society of New South Wales* 52, 161–184.

Medina, E., Garcia, V. and Cuevas, E. (1990). Sclerophylly and oligotrophic environments: relationships between leaf structure, mineral nutrient content, and drought resistance in tropical rain forest of the upper Rio Negro region. *Biotropica* 22, 51–64.

Megirian, D. (1992). Interpretation of the Miocene Carl Creek limestone, north-western Queensland. *The Beagle, Records of the Northern Territory Museum of Arts and Sciences* 9, 219–248.

Melick, D.R. (1990a). Ecology of rainforest and sclerophyllous communities in the Mitchell River National Park, Gippsland, Victoria. *Proceedings of the Royal Society of Victoria* 102, 71–87.

Melick, D.R. (1990b). Regenerative succession of *Tristaniopsis laurina* and *Acmena smithii* in riparian warm temperate rainforest in Victoria, in relation to light and nutrient regimes. *Australian Journal of Botany* 38, 111–120.

Melick, D.R. and Ashton, D.H. (1991). The effects of natural disturbances on warm temperate rainforests in south-eastern Australia. *Australian Journal of Botany* 39, 1–30.

Miller, G.H., Magee, J.W. and Jull, A.J. (1997). Low latitude glacial cooling in the southern hemisphere from amino-acid racemization in emu eggshells. *Nature* 385, 241–244.

Moore, A.D. and Noble, I.R. (1990). An individualistic model of vegetation stand dynamics. *Journal of Environmental Management* 31, 61–81.

Morrow, P.A. and Fox, L.R. (1980). Effects of variation in *Eucalyptus* essential oil yield on insect growth and grazing damage. *Oecologia* 45, 209–219.

Mount, A.B. (1964). The interdependence of the eucalypts and forest fires in southern Australia. *Australian Forestry* **28**, 166–172.

Mount, A.B. (1979). Natural regeneration processes in Tasmania forests. *Search* **10**, 180–186.

Mount, A.B. (1982). Fire-cycles or succession in south-west Tasmania. *Search* **13**, 174–175.

Murray, P. and Megirian, D. (1992). Continuity and contrast in middle and late Miocene vertebrate communities from the Northern Territory. *The Beagle, Records of the Northern Territory Museum of Arts and Sciences* **9**, 195–218.

Mutch, R.W. (1970). Wildland fires and ecosystems—a hypothesis. *Ecology* **51**, 1046–1051.

Myers, B.A., Duff, G., Eamus, D., Fordyce, I., O'Grady, A. and Williams, R.J. (1997). Seasonal variation in water relations of trees of differing leaf phenology in a wet-dry tropical savanna near Darwin, northern Australia. *Australian Journal of Botany* **45**, 225–240.

Myers, B.J., Robichaux, R.H., Unwin, G.L. and Craig, I.E. (1987). Leaf water relations and anatomy of a tropical rainforest tree species vary with crown position. *Oecologia* **74**, 81–85.

Nagle, J. (1991). Assessment of a rainforest reforestation project at Victoria Park Nature Reserve northern New South Wales. In *Rainforest Remnants* (ed. S. Phillips) pp. 116–121. (NSW National Parks and Wildlife Service: Sydney.)

Needham, R.H. (1960). Problems associated with regeneration of *Eucalyptus gigantea* in the Surrey Hills area. *Appita* **13**, 136–140.

Neilsen, W.A. and Ellis, R.A. (1981). Slash burning on forest sites: further comment. *Search* **12**, 9–10.

Neilsen, W.A. and Palzer, C. (1977). Analysis of foliar samples of healthy and dieback *Eucalyptus obliqua*. *Australian Forestry* **40**, 219–222.

Nelson, E.C. (1981). Phytogeography of southern Australia. In *Ecological Biogeography of Australia* (ed. A. Keast) pp. 733–759. (Dr. W. Junk: The Hague.)

Nicholson, P.H. (1981). Fire and the Australian Aborigine—an enigma. In *Fire and the Australian biota* (eds. A.M. Gill, R.H. Groves and I.R. Noble) pp. 55–76. (Australian Academy of Sciences: Canberra.)

Nix, H.A. (1982). Environmental determinants of biogeography and evolution in Terra Australis. In *Evolution of the flora and fauna of arid Australia* (eds. W.R. Barker and P.J.M. Greenslade) pp. 47–66. (Peacock Publications: Adelaide.)

Nix, H.A. (1991). An environmental analysis of Australian rainforests. In *The rainforest legacy: Australian National Rainforests Study. Volume 2—Flora and fauna of the rainforests* (eds. G. Werren and P. Kershaw) pp. 1–26. (Australian Government Publishing Service: Canberra.)

Noble, I.R. and Gitay, H. (1996). A functional classification for predicting the dynamics of landscapes. *Journal of Vegetation Science* **7**, 329–336.

Noble, I.R. and Slatyer, R.O. (1980). The use of vital attributes to predict successional changes in plant communities subject to recurrent disturbance. *Vegetatio* **43**, 5–21.

Noble, I.R. and Slatyer, R.O. (1981). Concepts and models of succession in vascular plant communities subject to recurrent fire. Fire in tall open-forest (wet sclerophyll forests). In *Fire and the Australian biota* (eds. A.M. Gill, R.H. Groves and I.R. Noble) pp. 311–335. (Australian Academy of Sciences: Canberra.)

Nott, J. (1995). The antiquity of landscapes on the north Australian craton and the implications for theories of long-term landscape evolution. *Journal of Geology* 103, 19–32.

O'Connell, A.M., Grove, T.S. and Dimmock, G.M. (1978). Nutrients in the litter on jarrah forest soils. *Australian Journal of Ecology* 3, 253–260.

O'Connell, J.F. and Allen, J. (1998). When did humans first arrive in greater Australia and why is it important to know? *Evolutionary Anthropology* 6, 132–146.

Olesen, T. (1992). Daylight spectra (400–740 nm) beneath sunny, blue skies in Tasmania, and the effect of a forest canopy. *Australian Journal of Ecology* 17, 451–461.

Olesen, T. (1997). The relative shade-tolerance of *Atherosperma moschatum* and *Elaeocarpus holopetalus*. *Australian Journal of Ecology* 22, 113–116.

Osborn, T.G.B. and Robertson, R.N. (1939). A reconnaissance survey of the vegetation of the Myall Lakes. *Proceedings of the Linnean Society of New South Wales* 64, 279–296.

Osunkjoya, O.O., Ash, J.E., Hopkins, M.S. and Graham, A.W. (1992). Factors affecting survival of tree seedlings in North Queensland rainforest. *Oecologia* 91, 569–578.

Osunkoya, O.O. and Ash, J.E. (1991). Acclimation to a change in light regime in seedlings of six Australian rainforest tree species. *Australian Journal of Botany* 39, 591–605.

Osunkoya, O.O., Ash, J.E., Graham, A.W. and Hopkins, M.S. (1993). Growth of tree seedlings in tropical rain forests of North Queensland, Australia. *Journal of Tropical Ecology* 9, 1–18.

Panton, W.J. (1993). Changes in post World War II distribution and status of monsoon rainforests in the Darwin area. *Australian Geographer* 24 (2), 50–59.

Patton, R.T. (1933). Ecological studies in Victoria. II. The fern gully. *Proceedings of the Royal Society of Victoria* 46, 117–129.

Peace, W.J.H. and Macdonald, F.D. (1981). An investigation of the leaf anatomy, foliar mineral levels, and water relations of trees of a Sarawak forest. *Biotropica* 13, 100–109.

Pearce, R.H. and Barbetti, M. (1981). A 38,000-year-old archaeological site at Upper Swan, Western Australia. *Archaeology in Oceania* 16, 173–178.

Pearcy, R.W. (1987). Photosynthetic gas exchange responses of Australian tropical forest trees in canopy, gap and understorey micro-environments. *Functional Ecology* 1, 169–178.

Petrie, A.H.K., Jarrett, P.H. and Patton, R.T. (1929). The vegetation on the Blacks' Spur region. A study in the ecology of some Australian mountain *Eucalyptus* forests I. The mature plant communities. *Journal of Ecology* 17, 223–248.

Phillips, O. (1995). Evaluating turnover in tropical forests. Response to Shiel. *Science* 268, 894–895.

Phillips, O.L. and Gentry, A.H. (1994). Increasing turnover through time in tropical forests. Science 263, 954–958.

Pidgeon, I.M. (1937). The ecology of the central coastal area of New South Wales. I. The environment and general features of the vegetation. *Proceedings of the Linnean Society of New South Wales* 62, 315–340.

Pidgeon, I.M. (1940). The ecology of the central coast of New South Wales. III. Types of primary succession. *Proceedings of the Linnean Society of New South Wales* 65, 221–249.

Podger, F.D., Bird, T. and Brown M.J. (1988). Human activity, fire and change in the forest at Hogsback Plain, southern Tasmania. In *Australia's ever changing forests* (eds. K.J. Frawley and N.M. Semple) pp. 119–140. (Department of Geography and Oceanography, Australian Defence Force Academy: Canberra.)

Pole, M.S. (1993). Keeping in touch: vegetation prehistory on both sides of the Tasman. *Australian Systematic Botany* 6, 387–397.

Pole, M. (1998). The fossil flora of Melville Island, Northern Australia. *The Beagle, Records of the Museums and Art Galleries of the Northern Territory* 14, 1–28.

Pole, M.S. and Bowman, D.M.J.S. (1996). Tertiary plant fossils from Australia's 'Top End'. *Australian Systematic Botany* 9, 113–126.

Pompe, A. and Vines, R.G. (1966). The influence of moisture on the combustion of leaves. *Australian Forestry* 30, 231–241.

Possingham, H.P., Comins, H.N. and Noble, I.R. (1995). The fire and flammability niches in plant communities. *Journal of Theoretical Biology* 174, 97–108.

Potts, B.M. and Reid, J.B. (1990). The evolutionary significance of hybridization *in Eucalyptus*. *Evolution* 44, 2151–2152.

Pressland, A.J. (1976). Effect of stand density on water use of mulga (*Acacia aneura* F. Muell.) woodlands in south-western Queensland. *Australian Journal of Botany* 24, 177–191.

Price, O., Bach, C., Shapcott, A. and Palmer, C. (1998). *Design of reserves for mobile species in monsoon rainforest.* (Parks and Wildlife Commission of the Northern Territory: Darwin.)

Price O. and Bowman, D.M.J.S. (1994). Fire-stick forestry: a matrix model in support of skilful fire management of *Callitris intratropica* R.T. Baker by north Australian Aborigines. *Journal of Biogeography* 21, 573–580.

Pryor, L.D. (1956). Chlorosis and lack of vigor in seedlings of Renantherous species of *Eucalyptus* caused by lack of mycorrhiza. *Proceedings of the Linnean Society of New South Wales* 81, 91–96.

Pryor, L.D. (1976). *The biology of Eucalypts.* (Edward Arnold: London.)

Pryor, L.D. and Johnson, L.A.S. (1981). *Eucalyptus*, the universal Australian. In *Ecological biogeography of Australia* (ed. A. Keast) pp. 499–536. (Dr W. Junk: The Hague.)

Pulsford, I.F., Banks, J.C.G. and Hodges, L. (1993). Land use history of the white cypress pine forests in the Snowy Valley, Kosciusko National Park. In *Australia's ever-changing forests II.* (eds. J. Dargavel and S. Feary) pp. 85–104. (Centre for Resource and Environmental Studies, Australian National University: Canberra.)

Quilty, P.G. (1994). The background: 144 million years of Australian palaeoclimate and palaeogeography. In *History of the Australian vegetation: Cretaceous to Recent* (ed. R.S. Hill) pp. 14–43. (Cambridge University Press: Cambridge.)

Raison, R.J. (1980). Possible forest site deterioration associated with slash-burning. *Search* 11, 68–71.

Raven, P.H. and Axelrod, D.I. (1972). Plate tectonics and Australasian paleobiogeography. *Science* 176, 1379–1386.

Raymo, M.E. and Ruddiman, W.F. (1992). Tectonic forcing of late Cenozoic climate. *Nature* 359, 117–122.

Read, J. (1985). Photosynthetic and growth responses to different light regimes of the major canopy species of Tasmanian cool temperate rainforest. *Australian Journal of Ecology* 10, 327–334.

Read, J. (1990). Some effects of acclimation temperature on net photosynthesis in some tropical and extra-tropical Australasian *Nothofagus* species. *Journal of Ecology* 78, 100–112.

Read, J. (1995). The importance of comparative growth rates in determining the canopy composition of Tasmanian rainforest. *Australian Journal of Botany* 43, 243–271.

Read, J. and Busby, J.R. (1990). Comparative responses to temperature of the major canopy species of Tasmanian cool temperate rainforest and their ecological significance. II. Net photosynthesis and climate analysis. *Australian Journal of Botany* 38, 185–205.

Read, J. and Francis, J. (1992). Responses of some Southern Hemisphere tree species to a prolonged dark period and their implications for high-latitude Cretaceous and Tertiary floras. *Palaeogeography, Palaeoclimatology, Palaeoecology* 99, 271–290.

Read, J. and Hill, R.S. (1985). Photosynthetic responses to light of Australian and Chilean species of *Nothofagus* and their relevance to the rainforest dynamics. *New Phytologist* 101, 731–742.

Read, J. and Hill, R.S. (1988). Comparative responses to temperature of the major canopy species of Tasmanian cool temperate rainforest and their ecological significance. I. Foliar frost resistance. *Australian Journal of Botany* 36, 131–143.

Read, J. and Hill, R.S. (1989). The response of some Australian temperate rain forest tree species to freezing temperatures and its biogeographical significance. *Journal of Biogeography* 16, 21–27.

Read, J. and Hope, G.S. (1989). Foliar frost resistance of some evergreen tropical and extratropical Australasian *Nothofagus* species. *Australian Journal of Botany* 37, 361–373.

Reichel, H. and Andersen A.N. (1996). The rainforest ant fauna of Australia's Northern Territory. *Australian Journal of Zoology* 44, 81–95.

Renbuss, M.A., Chilvers, G.A. and Pryor, L.D. (1972). Microbiology of an ashbed. *Proceedings of the Linnean Society of New South Wales* 97, 302–310.

Retallack, G.J. (1992). Middle Miocene fossil plants from Fort Ternan (Kenya) and evolution of African grasslands. *Paleobiology* 18, 383–400.

Richards, B.N. (1967). Introduction of the rain-forest species *Araucaria cunninghamii* Ait. to a dry sclerophyll forest environment. *Plant and Soil* 27, 201–216.

Richards, B.N. (1968). Effects of soil fertility on the distribution of plant communities as shown by pot cultures and field trials. *Commonwealth Forestry Review* 47, 200–210.

Richards, B.N. and Bevege, D.I. (1969). Critical foliage concentrations of nitrogen and phosphorus as a guide to the nutrient status of *Araucaria* underplanted to *Pinus*. *Plant and Soil* 31, 328–336.

Richards, P.W. (1952). *The tropical rain forest: an ecological study*. (Cambridge University Press: London.)

Ritchie, R. (1989). *Seeing the rainforests in 19th-century Australia*. (Rainforest Publishing: Sydney.)

Ridley, W.F. and Gardner, A. (1961). Fires in rain forest. *Australian Journal of Science* 23, 227–228.

Roberts, R.G, Jones, R. and Smith M.A. (1990). Thermoluminescence dating of a 50,000-year-old human occupation site in northern Australia. *Nature* 345, 153–156.

Roberts, R.G., Jones, R., Spooner, N.A., Head, M.J., Murray, A.S. and Smith, M.A. (1994). The human colonisation of Australia: optical dates of 53,000 and 60,000 years brackets human arrival at Deaf Adder Gorge, Northern Territory. *Quaternary Geochronology (Quaternary Science Reviews)* 13, 575–583.

Robinson, J.M. (1989). Phanerozoic O_2 variation, fire, and terrestrial ecology. *Palaeogeography, Palaeoclimatology, Palaeoecology (Global Planetary Change Section)* 75, 223–240.

Robinson, J.M. (1994). Speculations on carbon dioxide starvation, Late Tertiary evolution of stomatal regulation and floristic modernization. *Plant, Cell and Environment* 17, 345–354.

Roche, S., Dixon, K.W. and Pate, J.S. (1997a). Seed ageing and smoke: partner cures in the amelioration of seed dormancy in selected Australian native species. *Australian Journal of Botany* 45, 783–815.

Roche, S., Dixon, K.W. and Pate, J.S. (1998). For everything a season: smoke-induced seed germination and seedling recruitment in a Western Australian *Banksia* woodland. *Australian Journal of Ecology* 23, 111–120.

Roche, S., Koch, J.M. and Dixon, K.W. (1997b). Smoke enhanced seed germination for mine rehabilitation in the southwest of Western Australia. *Restoration Ecology* 5, 191–203.

Rozefelds, A.C. (1996). *Eucalyptus* phylogeny and history: a brief summary. *Tasforests* 8, 15–26.

Russell-Smith, J. (1985a). A record of change: studies of Holocene vegetation history in the South Alligator River region, Northern Territory. *Proceedings of the Ecological Society of Australia* 13, 191–202.

Russell-Smith, J. (1985b). Studies in the jungle: people, fire and monsoon forest. In *Archaeological Research in Kakadu National Park* (ed. R. Jones) pp. 241–267. (Australian National Parks and Wildlife Service Special Publication No. 13: Canberra.)

Russell-Smith, J. (1991). Classification, species richness, and environmental relations of monsoon rain forest in northern Australia. *Journal of Vegetation Science* 2, 259–278.

Russell-Smith, J. (1996). Regeneration of monsoon rain forest in northern Australia: the sapling bank. *Journal of Vegetation Science* 7, 889–900.

Russell-Smith, J. and Bowman, D.M.J.S. (1992). Conservation of monsoon rainforest isolates in the Northern Territory, Australia. *Biological Conservation* 59, 51–63.

Russell-Smith, J. and Lee, A.H. (1992). Plant populations and monsoon rain forest in the Northern Territory, Australia. *Biotropica* 24, 471–487.

Russell-Smith, J. and Lucas, D.E. (1994). Regeneration of monsoon rain forest in northern Australia: the dormant seed bank. *Journal of Vegetation Science* 5, 161–168.

Russell-Smith, J., Lucas, D.E., Brock, J. and Bowman, D.M.J.S. (1993). *Allosyncarpia*-dominated rain forest in monsoonal northern Australia. *Journal of Vegetation Science* 4, 67–82.

Russell-Smith, J., Lucas, D., Gapindi, M., Gunbunuka, B., Kapirigi, N., Namingum, G., Lucas, K., Giuliani, P. and Chaloupka, G. (1997a). Aboriginal resource utilization and fire management practice in western Arnhem Land, monsoonal northern Australia: notes for prehistory, lessons for the future. *Human Ecology* 25, 159–195.

Russell-Smith, J., Ryan, P.G. and Durieu, R. (1997b). A LANDSAT MSS-derived fire history of Kakadu National Park, monsoonal northern Australia, 1980–94: seasonal extent, frequency and patchiness. *Journal of Applied Ecology* 34, 748–766.

Sakai, A., Paton, D.M. and Wardle, P. (1981). Freezing resistance of trees of the south temperate zone, especially subalpine species of Australasia. *Ecology* 62, 563–570.

Schimper, A.F.W. (1903). *Plant-geography upon a Physiological Basis* (translated by W.R. Fisher, revised and edited by P. Groom and I.B. Balfour). (Oxford University Press: Oxford.)

Seddon, G. (1974). Xerophytes, xeromorphs and sclerophylls: the history of some concepts in ecology. *Biological Journal of the Linnean Society* 6, 65–87.

Seddon, G. (1984). Characteristics and classification of rainforest. *Landscape Australia* 4/84, 276–285.

Seddon, G. (1985). The conservation of rainforest. *Landscape Australia* 1/85, 20–31.

Setterfield, S.A. and Williams, R.J. (1996). Patterns of flowering and seed production in *Eucalyptus miniata* and *E. tetrodonta* in a tropical savanna woodland, northern Australia. *Australian Journal of Botany* **44**, 107–122.

Shackleton, N.J. and Opdyke, N.D. (1973). Oxygen isotope and palaeomagnetic stratigraphy of equatorial Pacific Core V28-238: oxygen isotope temperatures and ice volumes on a 10^5 year and 10^6 year scale. *Quaternary Research* **3**, 39–55.

Sheil, D. (1995). Evaluating turnover in tropical forests. *Science* **268**, 894.

Shulmeister J. (1992). A Holocene pollen record from lowland tropical Australia. *Holocene* **2**, 107–116.

Singh, G. (1988). History of aridland vegetation and climate: a global perspective. *Biological Review* **63**, 159–195.

Singh, G. and Geissler, E.A. (1985). Late Cainozoic history of vegetation, fire, lake levels and climate, at Lake George, New South Wales, Australia. *Philosophical Transactions of the Royal Society of London B* **311**, 379–447.

Singh, G., Kershaw, A.P. and Clark R. (1981). Quaternary vegetation and fire history in Australia. In *Fire and the Australian biota* (eds. A.M Gill, R.H. Groves, and I.R. Noble) pp. 23–54. (Australian Academy of Science: Canberra.)

Singh, G. and Luly, J. (1991). Changes in vegetation and seasonal climate since the last full glacial at Lake Frome, South Australia. *Palaeogeography, Palaeoclimatology, Palaeoecology* **84**, 75–86.

Sluiter, I.R.K. (1992). The nature of the Tertiary flora and its legacy in modern Victorian vegetation. In *Victoria's rainforests: perspectives on definition, classification and management* (eds. P. Gell and D. Mercer) pp. 97–105. (Monash Publications in Geography No. 41: Melbourne.)

Sluiter, I.R. and Kershaw, A.P. (1982). The nature of Late Tertiary vegetation in Australia. *Alcheringa* **6**, 211–222.

Smith, A.P and Ganzhorn, J.U. (1996). Convergence in community structure and dietary adaptation in Australian possums and gliders and Malagasy lemurs. *Australian Journal of Ecology* **21**, 31–46.

Smith, J.M.B. and Guyer, I.J. (1983). Rainforest–eucalypt forest interactions and the relevance of the biological nomad concept. *Australian Journal of Ecology* **8**, 55–60.

Smith, M.A. (1987). Pleistocene occupation in arid Central Australia. *Nature* **328**, 710–711.

Snyder, J.R. (1984). The role of fire: Mutch ado about nothing? *Oikos* **43**, 404–405.

Specht, R.L. (1957). Dark Island heath (Ninety-Mile Plain, South Australia). *Australian Journal of Botany* **5**, 151–172.

Specht, R.L. (1958). The climate, geology, soils and plant ecology of the northern portion of Arnhem Land. In *Records of the American-Australian Scientific Expedition to Arnhem Land Volume 3 Botany and plant ecology* (eds. R.L. Specht and C.P. Mountford) pp. 333–414. (Melbourne University Press: Melbourne.)

Specht, R.L. (1972). Water use by perennial evergreen plant communities in Australia and Papua New Guinea. *Australian Journal of Botany* 20, 273–299.

Specht, R.L. (1981*a*). Foliage projective cover and standing biomass. In *Vegetation Classification in Australia* (eds. A.N. Gillison and D.J. Anderson) pp. 10–21. (Commonwealth Scientific and Industrial Research Organisation and Australian National University: Canberra.)

Specht, R.L. (1981*b*). Ecophysiological principles determining the biogeography of major vegetation formations in Australia. In *Ecological biogeography of Australia* (ed. A. Keast) pp. 299–333. (Dr W. Junk: The Hague.)

Specht, R.L. (1981*c*). Growth indices—their role in understanding the growth, structure and distribution of Australian vegetation. *Oecologia* 50, 347–356.

Specht, R.L. (1983). Foliage projective covers of overstorey and understorey strata of mature vegetation in Australia. *Australian Journal of Ecology* 8, 433–439.

Specht, R.L. and Groves, R.H. (1966). A comparison of the phosphorus nutrition of Australian heath plants and introduced economic plants. *Australian Journal of Botany* 14, 201–221.

Specht, R.L. and Jones, R. (1971). A comparison of the water use by heath vegetation at Frankston, Victoria, and Dark Island Soak, South Australia. *Australian Journal of Botany* 19, 311–326.

Specht, R.L. and Rundel, P.W. (1990). Sclerophylly and foliar nutrient status of mediterranean-climate plant communities in southern Australia. *Australian Journal of Botany* 38, 459–474.

Specht, R.L., Salt, R.B. and Reynolds, S.T. (1977). Vegetation in the vicinity of Weipa, North Queensland. *Proceedings of the Royal Society of Queensland* 88, 17–38.

Spooner, N.A. (1998). Human occupation at Jinmium, northern Australia: 116,000 years ago or much less? *Antiquity* 72, 173–178.

Stocker, G.C. (1969). Fertility differences between the surface soils of monsoon and eucalypt forests in the Northern Territory. *Australian Forest Research* 4, 31–38.

Stocker G.C. (1971). The age of charcoal from old jungle fowl nests and vegetation change on Melville Island. *Search* 2, 28–30.

Stocker, G.C. (1981). Regeneration of a north Queensland rain forest following felling and burning. *Biotropica* 13, 86–92.

Stocker, G.C. and Mott, J.J. (1981). Fire in the tropical forests and woodlands of northern Australia. In *Fire and the Australian biota* (eds. A.M. Gill, R.H. Groves and I.R. Noble) pp. 425–439. (Australian Academy of Science: Canberra.)

Stockton, J. (1982). Fires by the seaside: historic vegetation changes in north-western Tasmania. *Papers and Proceedings of the Royal Society of Tasmania* 116, 53–66.

Taylor, G. (1994). Landscapes of Australia: their nature and evolution. In *History of the Australian vegetation: Cretaceous to Recent* (ed. R.S. Hill) pp. 60–79. (Cambridge University Press: Cambridge.)

Thomas I. (1993). Late Pleistocene environments and Aboriginal settlement patterns in Tasmania. *Australian Archaeology* 36, 1–11.

Thomas I. (1995). Models and prima-donnas in southwest Tasmania. *Australian Archaeology* 41, 21–23.

Thompson, C.H. and Bowman, G.M. (1984). Subaerial denudation and weathering of vegetated coastal dunes in eastern Australia. In *Coastal Geomorphology in Australia* (ed. B.G. Thom) pp. 263–290. (Academic Press: Sydney.)

Thompson, W.A., Stocker, G.C. and Kriedemann, P.E. (1988). Growth and photosynthetic response to light and nutrients of *Flindersia brayleyana* F. Muell., a rainforest tree with broad tolerance to sun and shade. *Australian Journal of Plant Physiology* 15, 299–315.

Thomson, D.F. (1939). The seasonal factor in human culture: illustrated from the life of a contemporary nomadic group. *Proceedings of the Prehistoric Society* 5, 209–221.

Thomson, D.F. (1949). Arnhem Land: explorations among an unknown people II. The people of Blue Mud Bay. *Geographical Journal* 63, 1–8.

Tommerup, E.C. (1934). Plant ecological studies in south-east Queensland. *Proceedings of the Royal Society of Queensland* 46, 91–118.

Tracey, J.G. (1969). Edaphic differentiation of some forest types in eastern Australia. I. Soil physical factors. *Journal of Ecology* 57, 805–816.

Troumbis, A.Y. and Trabaud, L. (1989). Some questions about flammability in fire ecology. *Acta Ecologica/Ecologia Plantarum* 10, 167–175.

Truswell, E.M. (1993). Vegetation changes in the Australian Tertiary in response to climatic and phytogeographic forcing factors. *Australian Systematic Botany* 6, 533–557.

Truswell, E.M. and Harris, W.K. (1982). The Cainozoic palaeobotanical record in arid Australia: fossil evidence for the origins of an arid-adapted flora. In *Evolution of the flora and fauna of arid Australia* (eds. W.R. Barker and P.J.M. Greenslade) pp. 67–76. (Peacock Publications: Adelaide.)

Tunstall, B.R. and Connor, D.J. (1975). Internal water balance of brigalow (*Acacia harpophylla* F. Muell.) under natural conditions. *Australian Journal of Plant Physiology* 2, 489–499.

Tunstall, B.R. and Connor, D.J. (1981). A hydrological study of a subtropical semiarid forest of *Acacia harpophylla* F. Muell. ex Benth. (Brigalow). *Australian Journal of Botany* 29, 311–320.

Turnbull, M.H. (1991). The effect of light quantity and quality during development on the photosynthetic characteristics of six Australian rainforest tree species. *Oecologia* 87, 110–117.

Turner, I.M. (1994a). A quantitative analysis of leaf form in woody plants from the world's major broadleaved forest types. *Journal of Biogeography* 21, 413–419.

Turner, I.M. (1994b). Sclerophylly: primarily protective? *Functional Ecology* 8, 669–675.

Turner, J.C. (1976). An attitudinal transect in rain forest in the Barrington Tops area, New South Wales. *Australian Journal of Ecology* 1, 155–174.

Turner, J. and Kelly, J. (1981). Relationships between soil nutrients and vegetation in a north coast forest, New South Wales. *Australian Forestry Research* 11, 201–208.

Turton, S.M. (1990). Light environments within montane tropical rainforest, Mt Bellenden Ker, North Queensland. *Australian Journal of Ecology* 15, 35–42.

Turton, S.M. (1992). Understorey light environments in a north-east Australian rain forest before and after a tropical cyclone. *Journal of Tropical Ecology* 8, 241–252.

Turton, S.M. and Duff, G.A. (1992). Light environments and floristic composition across an open forest-rainforest boundary in northeastern Queensland. *Australian Journal of Ecology* 17, 415–423.

Turton, S.M. and Sexton, G.J. (1996). Environmental gradients across four rainforest-open forest boundaries in northeastern Queensland. *Australian Journal of Ecology* 21, 245–254.

Udovicic, F., McFadden, G.I. and Ladiges, P.Y. (1995). Phylogeny of *Eucalyptus* and *Angophora* based on 5S rDNA spacer sequence data. *Molecular Phylogenetics and Evolution* 4, 247–256.

UNESCO (1973). *International classification and mapping of vegetation.* (UNESCO: Paris.)

Unwin, G.L. (1989). Structure and composition of the abrupt rainforest boundary in the Herberton Highland, north Queensland. *Australian Journal of Botany* 37, 413–428.

Unwin, G.L., Applegate, G.B., Stocker, G.C. and Nicholson, D.I. (1988). Initial effects of tropical cyclone 'Winifred' on forests in north Queensland. *Proceedings of the Ecological Society of Australia* 15, 283–296.

Unwin, G.L. and Kriedemann, P.E. (1990). Drought tolerance and rainforest tree growth on a north Queensland rainfall gradient. *Forest Ecology and Management* 30, 113–123.

Unwin, G.L., Stocker, G.C, and Sanderson, K.D. (1985). Fire and the forest ecotone in the Herberton highland, north Queensland. *Proceeding of the Ecological Society of Australia* 13, 215–224.

Vernes, K., Marsh, H. and Winter, J. (1995). Home-range characteristics and movement patterns of the red-legged pademelon (*Thylogale stigmatica*) in a fragmented tropical rainforest. *Wildlife Research* 22, 699–708.

Vitousek, P.M. (1984). Litterfall, nutrient cycling and nutrient limitation in tropical forests. *Ecology* 65, 285–298.

Walker, D. and Chen Y. (1987). Palynological light on tropical rainforest dynamics. *Quaternary Science Reviews* 6, 77–92.

Walker, J. and Hopkins, M.S. (1984). Vegetation. In *Australian soil and land survey field handbook* pp. 44–67. (Inkata Press: Melbourne.)

Wardell-Johnston, G.W., Williams, J.E., Hill, K.D. and Cumming, R. (1997). Evolutionary biogeography and contemporary distribution of eucalypts. In *Eucalypt ecology: individuals to ecosystems* (eds. J.E. Williams and J.C.Z. Woinarski) pp. 92–128. (Cambridge University Press: Cambridge.)

Wasson, R.J. (1986). Geomorphology and Quaternary history of the Australian continental dunefields. *Geographical Review of Japan* 59, 55–67.

Webb, L.J. (1954). Aluminium accumulation in the Australian-New Guinea flora. *Australian Journal of Botany* 2, 176–196.

Webb, L.J. (1958). Cyclones as an ecological factor in tropical lowland rain-forest, north Queensland. *Australian Journal of Botany* 6, 220–228.

Webb, L.J. (1959). A physiognomic classification of Australian rain forests. *Journal of Ecology* 47, 551–570.

Webb, L.J. (1960). A new attempt to classify Australian rain forests. *Silva Fennica* 105, 98–104.

Webb, L.J. (1964). An historical interpretation of the grass balds of the Bunya Mountains, south Queensland. *Ecology* 45, 159–162.

Webb, L.J. (1968). Environmental relationships of the structural types of Australian rain forest vegetation. *Ecology* 49, 296–311.

Webb, L.J. (1969). Edaphic differentiation of some forest types in eastern Australia II. Soil chemical factors. *Journal of Ecology* 57, 817–830.

Webb, L.J. (1977). Ethnobotany: the co-operative approach to research. *Australian Institute of Aboriginal Studies Newsletter* 7, 43–45.

Webb, L.J. (1978). A general classification of Australian rainforests. *Australian Plants* 9, 349–363.

Webb, L.J. (1983). Ecological values of the tropical rainforest resource. *Proceedings of the Linnean Society of New South Wales* 106, 263–274.

Webb, L. (1992). The case for an ecologically-based definition of Australia's rainforests. In *Victoria's rainforests: perspectives on definition, classification and management* (eds. P. Gell and D. Mercer) pp. 171–177. (Monash Publications in Geography No. 41: Melbourne.)

Webb, L.J. and Tracey, J.G. (1981). Australian rainforests: patterns and change. In *Ecological Biogeography of Australia* (ed. A. Keast) pp. 605–694. (Dr. W. Junk: The Hague.)

Webb, L.J., Tracey, J.G. and Jessup, L.W. (1986). Recent evidence for autochthony of Australian tropical and subtropical rainforest floristic elements. *Telopea* 2, 575–589.

Webb, L.J., Tracey, J.G. and Williams, W.T. (1976). The value of structural features in tropical forest typology. *Australian Journal of Ecology* 1, 3–28.

Webb, L.J., Tracey, J.G. and Williams, W.T. (1984). A floristic framework of Australian rainforests. *Australian Journal of Ecology* 9, 169–198.

Webb, L.J., Tracey, J.G., Williams, W.T. and Lance, G.N. (1969). The pattern of mineral return in leaf litter of three subtropical Australian forests. *Australian Forestry* 33, 99–110.

Westman, W.E. (1990). Structural and floristic attributes of recolonizing species in large rain forest gaps, north Queensland. *Biotropica* 22, 226–234.

Westman, W.E. and Rogers, R.W. (1977). Nutrient stocks in a subtropical eucalypt forest, North Stradbroke Island. *Australian Journal of Ecology* 2, 447–460.

Weston, P.H. and Crisp, M.D. (1994). Cladistic biogeography of Waratahs

(Proteaceae: Embothrieae) and their allies across the Pacific. *Australian Systematic Botany* 7, 225–249.

Wharton C.H. (1969). Man, fire and wild cattle in Southeast Asia. *Tall Timbers Fire Conference* 8, 107–167.

White, J.P. (1994). Site 820 and the evidence for early occupation in Australia. *Quaternary Australasia* 12, 21–23.

Whitmore, T.C. (1990). *An Introduction to Tropical Rain Forests.* (Clarendon Press: Oxford).

Wilford, G.E. and Brown, P.J. (1994). Maps of late Mesozoic-Cenozoic Gondwana break-up: some palaeogeographical implications. In *History of the Australian vegetation: Cretaceous to Recent* (ed. R.S. Hill) pp. 5–13. (Cambridge University Press: Cambridge.)

Williams J.E. and Woinarski, J.C.Z. (1997). *Eucalypt ecology: individuals to ecosystems.* (Cambridge University Press: Cambridge.)

Williams, K.J. and Potts, B.M. (1996). The natural distribution of *Eucalyptus* species in Tasmania. *Tasforests* 8, 39–165.

Winter, J.W., Atherton, R.G., Bell, F.C. and Pahl, L.I. (1987). An introduction to Australian rainforests. In *The rainforest legacy: Australian national rainforests study. Volume I—The nature, distribution and status of rainforest types* (eds G.L. Werren and A.P. Kershaw) pp. 1–7. (Australian Government Publishing Service: Canberra.)

Withers, J.R. (1979). Studies on the status of unburnt *Eucalyptus* woodland at Ocean Grove, Victoria. IV The effect of shading on seedling establishment. *Australian Journal of Botany* 27, 47–66.

Wood, J.G. (1959). The phytogeography of Australia (in relation to radiation of *Eucalyptus, Acacia,* etc). In *Biogeography and ecology of Australia* (eds. J.A. Keast, R.L. Crocker and C.S. Christian) pp. 291–302. (Dr. W. Junk: The Hague.)

Wood, J.G. and Williams, R.J. (1960). Vegetation. In *The Australian Environment* (ed. G.W. Leeper) pp. 67–84. (Commonwealth Scientific and Industrial Research Organisation and Melbourne University Press: Melbourne.)

Woodroffe, C.D. and Chappell, J. (1993). Holocene emergence and evolution of the McArthur River Delta, southwestern Gulf of Carpentaria, Australia. *Sedimentary Geology* 83, 303–317.

Woodroffe, C.D., Mulrennan, M.E. and Chappell, J. (1993). Estuarine infill and coastal progradation, southern van Diemen Gulf, northern Australia. *Sedimentary Geology* 83, 257–275.

Woodroffe, C.D., Thom, B.G. and Chappell, J. (1985). Development of widespread mangrove swamps in mid-Holocene times in northern Australia. *Nature* 317, 711–713.

Wright, R. (1986). How old is zone F at Lake George? *Archaeology in Oceania* 21, 138–139.

Yates, D.J. and Hutley, L.B. (1995). Foliar uptake of water by wet leaves of *Sloanea woolsii*, an Australian subtropical rainforest tree. *Australian Journal of Botany* **43**, 157–167.

Yates, D.J., Unwin, G.L. and Doley, D. (1988). Rainforest environment and physiology. *Proceedings of the Ecological Society of Australia* **15**, 31–37.

Young, R. and McDougall, I. (1993). Long-term landscape evolution: early Miocene and modern rivers in southern New South Wales, Australia. *Journal of Geology* **101**, 35–49.

Zedler, P.H. (1995). Are some plants born to burn? *Trends in Ecology and Evolution* **10**, 393–395.

Index

Note: Page references in normal font indicate text references, those in italics indicate entire chapters and those in bold indicate a figure or table

and soil structure 85, **87**
leaves 53, 57, **58**, **61**, 62, 94
nutrient cycling 187, **188**
tolerance of cold 151
charcoal
in pollen cores 46, 228, **229**, 230,
231, **232**, 232, 233–5, **235**, 237–9,
238, 241, 266–7, **266**
in soil 179, 232, 235–6, **237**
chenier plains 165, **165**
Chenopodiaceae **238**
Cissus 71
Citrus 71
classification of rainforest 14, **15**,
26–44
Baur's system 29–30
Beadle and Costin's system 28–9
by climate 14, 26–7
by floristics 14, **15**
by structure **38–9**
Webb's system 32, **33–7**, 47
see also definition of rainforest
clearing
of rainforest 2, 108, **113**, 275, 280,
281
of vegetation 11, 227
Clematis 71
Clements' succession theory 41, 199
Clerodendrum 71
Clerodendrum cunninghamii **259**
climate change
and fire 265–7, **266**, 275
in Cainozoic Australia 250–65,
252, 263, **266**
climatic limitation of rainforest 26,
99–133, *134–55*, 265
summer rainfall 104–5, 262, **262**,
263, 275
Clitoria temata 281
'closed forest' 32–40
coachwood *see Ceratopetalum
apetalum*
Codonocarpus 71
Commersonia 71

Complex mesophyll vine forest 33,
262
Complex notophyll vine forest
(CNVF) **35**, **139**, 262
conifers 43, 48, **140**, 151, 253
see also named families and genera
conservation of rainforest 278,
279–84
research for 283–4
see also Aborigines; mixed forest
continental drift 4, 251–2, **252**,
259–60
cool temperate rainforest *see* temperate
rainforest
Corchorus 71
Corymbia see Eucalyptus
Cretaceous Period 4, **244**
Cryptostegia grandiflora 279
Culcita **238**
Cunoniaceae **235**
Cyathea **238**
Cyathea robertsonii **238**
cyclones
and fire in rainforests 160, 162,
170, **171**, 172, 228–9, 280,
281–2
defoliation of rainforest 138, **140**
occurrence 4–5
Cynanchum 71
Cyperaceae **238**

Dacrycarpus **238**
Dacrydium 234, **235**, 237
Dacrydium cupressinum **238**
Dacrydium guillauminii **238**
Darlingia darlingiana 141
Darwin crater pollen core 239–40,
242
Davalliaceae **238**
deciduous beech *see Nothofagus gunnii*
Deciduous microphyll vine thicket
262
Deciduous vine thicket 37, 162, **262**,
265

335 Index